ROUTLEDGE LIBRARY EDITIONS: LOGIC

Volume 16

THE DEVELOPMENT OF MATHEMATICAL LOGIC

THE DEVELOPMENT OF
MATHEMATICAL LOGIC

P. H. NIDDITCH

LONDON AND NEW YORK

First published in 1962 by Routledge & Kegan Paul Ltd

This edition first published in 2020
by Routledge
2 Park Square, Milton Park, Abingdon, Oxon OX14 4RN

and by Routledge
52 Vanderbilt Avenue, New York, NY 10017

Routledge is an imprint of the Taylor & Francis Group, an informa business

© 1962 P. H. Nidditch

British Library Cataloguing in Publication Data
A catalogue record for this book is available from the British Library

ISBN: 978-0-367-41707-9 (Set)
ISBN: 978-0-367-81582-0 (Set) (ebk)
ISBN: 978-0-367-42071-0 (Volume 16) (hbk)
ISBN: 978-0-367-42624-8 (Volume 16) (pbk)
ISBN: 978-0-367-85421-8 (Volume 16) (ebk)

Publisher's Note
The publisher has gone to great lengths to ensure the quality of this reprint but points out that some imperfections in the original copies may be apparent.

Disclaimer
The publisher has made every effort to trace copyright holders and would welcome correspondence from those they have been unable to trace.

THE DEVELOPMENT OF
MATHEMATICAL LOGIC

BY

P. H. Nidditch

Routledge & Kegan Paul Ltd.
LONDON AND BOSTON

First published 1962
in Great Britain by
Routledge & Kegan Paul Ltd
Broadway House, 68–74 Carter Lane
London EC4V 5EL
and 9 *Park Street*
Boston, Mass. 02108, *U.S.A.*

Second impression 1963
Third impression 1966
Fourth impression 1972

ISBN 0 7100 7336 4 (*c*)

Reprinted by Photolithography in Great Britain by
Lowe & Brydone (Printers) Ltd., London

CONTENTS

Contents

Contents

Contents

1

PURPOSE AND LANGUAGE OF THE BOOK

1.1. Purpose of book. The purpose of the present book is to give such an account of Mathematical Logic as will make clear in the framework of its history some of the chief directions of its ideas and teachings. It is these directions, not the mass of detail forming the theory and its history, which are important for the rest of philosophy and are important, in addition, from the point of view of a general education. In the limits of our space we are able to give attention only to what has been, or seems as if it may be, fertile or of special value in some other way. More than this, Mathematical Logic having no small number of important developments, a selection of material is necessary, the selection being guided by the rule to take up the simpler questions, other things being roughly equal. Facts have to be looked at in the light of one's purpose. Though they may all have the same value simply as facts, they are not at all equal as judged by the profit and the pleasure that thought, and not least the thought of the learner, is able to get from them.

1.2. Language of book. The writing of all this discussion of Mathematical Logic is in Ogden's Basic English. One is forced when keeping to the apparatus of this form of the English language (which is a body of only 850 root words—not taking into account 51

1

international words, and names of numbers, weights and measures, of sciences and of other branches of learning, and up to 50 words for any field of science or learning, all of which words may be used in addition to the 850—and, in comparison with those of normal English, a very small number of rules limiting greatly the uses of the given words and the structures of statements that may be formed from them), to take more care than one commonly does to make the dark and complex thoughts that are at the back of one's mind as clear and simple as possible. We will be attempting in what is to come to get across to the reader the substance of the story of Mathematical Logic. In this attempt Basic English will certainly be of some help to some, possibly even to almost all, readers, and will certainly not make things harder for any.

All the special words of logic which are not among the 850 words of Basic English are put in sloping print when they are first made use of here and their senses are made clear by Basic English.

However, though we have done our best, by using Basic English, to keep the account as clear and simple as possible, it will not be surprising if there are bits of it which are hard fully to get a grip of in a first reading; this is specially true when the newer developments are being quickly talked about (9.1 to the end of the book). The reason for this is that the tendency of the growth of all sciences is towards the more and more complex and towards an ever-increasing number of high-level ideas in their organization. Certainly the growth of Mathematical Logic has been like this. But our hope is to have said nothing that will not be completely clear after some thought—thought will sometimes be needed. No earlier knowledge of logic will in any degree be necessary and only such a knowledge of high school mathematics as the reader still has in the mists of his memory.

2

ARISTOTLE'S SYLLOGISTIC

2.1. Mathematical Logic produced by four lines of thought. Mathematical Logic is the outcome of the joining together of four different lines of thought. These are the old logic, the invention of Aristotle; the idea of a complete and automatic language for reasoning; the new developments in algebra and geometry which took place after 1825; and the idea of the parts of mathematics as being systems of *deductions*, that is of chains of reasoning in agreement with rules of logic, these rules giving one the power to go from a statement s_1 to another statement s_2 when s_2 is necessarily true if s_1 is taken to be true. We will say something about these four in turn.

2.2. Aristotle's work on logic. Aristotle (384–322 B.C.) first, then. There are at least five sides to Aristotle's writings on logic. In these writings are: discussions of common language, chiefly in relation to the different sorts of words and their connection with the possible orders of existence (substance, quality, place, time, and so on); a body of suggestions on the art of argument—the art of causing the destruction of the arguments of those who are not in agreement with oneself and of stopping one's arguments from being open to like attacks; a group of teachings on the way of

3

science, on how an increase of knowledge of physical laws may come about by the work of the natural sciences; a number of views on the right organization of a system in the science of mathematics; and a theory of that form of certain reasoning which was named by Aristotle *syllogistic* reasoning. It is this last theory and its later offspring that one normally has in mind when talking of 'the old logic', 'the common logic' or 'the logic of Aristotle', and it is this side of his writings on logic that was important for the start and early growth of Mathematical Logic; so in what we now say about Aristotle we will be limiting ourselves to his Syllogistic.

2.3. Reasoning, implication and validity. Complex statements such as 'if all animals are in need of food and all men are animals, then all men are in need of food' and 'if the sides joining any three points in a plane make a right-angle at one of the points, then the measure of the square on the side opposite the right-angle is equal to the measure got by the addition of the measures of the squares on the other two sides' are the sort of statements put forward as examples of reasoning that is certain: reasoning which is to the effect that a given statement is certainly true if a number of other given statements offered as conditions for it are true. There is no suggestion that all reasoning put forward as certain is in fact certain; it is common knowledge that errors of reasoning are frequently made.

Now, more generally, let $s_1, s_2, \ldots, s_n$ be any n statements. In the logic of certain, as against for example probable, reasoning, a statement of the form 'if s_1 and s_2 and $\ldots$ and s_{n-1}, then necessarily s_n' is said to be an *implication*. ('Implication' by itself, without 'an' or 'the', is used as a name for the 'if-then necessarily' relation.)

4

s_1 to s_{n-1} are the conditions and s_n is the outcome of the implication. An implication is said to have *validity* if and only if it is necessary for the outcome to be true when all the conditions are true. The reader is not to let the idea of an implication be mixed with the idea of an argument. An argument is a complex statement of the form 's_1 is true and s_2 is true and . . . and s_{n-1} is true, so necessarily s_n is true' which has validity if and only if the parallel implication has validity. An argument is different from an implication in so far as the conditions on which the outcome is dependent are judged to be true and because of this the outcome is put forward as being true; in an implication, on the other hand, the outcome is put forward as being true if the conditions are true, but these are not judged to be in fact true, or false. ('Argument' is used, in addition, for any chain of reasoning; this might be an implication or it might be an argument in the narrower sense.)

2.4. General statements. In Aristotle's logic only four sorts of statement may be used as conditions or outcomes of implications. They are general statements, in a special sense of 'general', which have the structure 'all S is P', 'no S is P', 'some S is P' or 'some S is not P'; later, those of the first sort were said to be A, those of the second sort E, those of the third sort I and those of the fourth sort O statements, while for all of them S was named the *subject* and P the *predicate*. The way of Aristotle himself in talking about these four forms of statement was somewhat different from that which was made common by the Schoolmen. For him 'P is truly said of all S' or 'P is a part of whatever is S' took the place of 'all S is P', 'P is truly said of no S' or 'P is a part of nothing that is S' took the place of 'no S is P', and so on. In Aristotle's opinion the only names which make sense in a general statement are general names, for example 'man' and 'flower' and 'green', not names of persons or places or of other things that might be viewed as if they were units of being. In harmony with this opinion Aristotle did not let statements such as 'Socrates is going to be dead some day' and 'Callias has a turned-up nose'—examples he frequently makes use of in his writings on philosophy—be conditions or outcomes of implications.

It is only right to say in addition that Aristotle gave *modal* general statements and syllogistic reasonings made of these an important place in his chief work on Syllogistic, the *Prior Analytics*; such modal statements are ones about the necessary or the possible, for example 'all S is necessarily P', 'no S is necessarily P' and 'some S is possibly P'. Aristotle's modal logic had little effect on later thought and no effect, it seems, on the birth of Mathematical Logic. We will give no more attention to it.

2.5. Aristotle's Syllogistic. Aristotle's Syllogistic is a theory of syllogistic implications. A syllogistic implication is an implication with two and only two conditions and one outcome, the conditions and the outcome being general statements. What Aristotle made the attempt to do in his Syllogistic was to give a complete account of the different possible detailed forms of syllogistic implications and a complete body of rules as tests of the validity of any given syllogistic implication. Being a first attempt, it is not surprising that Aristotle's theory is not free from error and is not the best possible one. But, without any doubt, he made a solid start. It is sad that for hundreds of years after his ideas on Syllogistic had got a wide distribution among those

Aristotle's Syllogistic

with an interest in the theory of reasoning, almost no one was able to let himself say outright in what ways Aristotle's answers were right or wrong and, much more important, say that the questions they were designed as answers to were only some of the questions needing to be answered. It is true that some specially among the later Schoolmen did make important discoveries in logic and that their work here was not limited to Syllogistic. However, their logic was one of rules whose statement was in everyday language (Latin); no special signs for the operations of reasoning were used and the Schoolmen seem to have had no idea that it was possible for logic to be turned into a sort of mathematics. So they took no step forward in the direction of turning logic into Mathematical Logic. It would certainly have been hard for them to take seriously the idea of logic as a sort of mathematics because their view of logic and mathematics kept these quite separate. Logic (*ars logica*) was one of the three Arts of Language forming the *trivium* ('the three-branched way') which was the field of learning that one had to go through first when at the university; the other two branches of the *trivium* were the art of using language rightly (*ars grammatica*) and the art of using language well (*ars rhetorica*). The teaching of the *trivium* was done on the Arts side of the universities. On the other hand, the business of mathematics was not seen as being with language at all and its teaching was given on the Science side of the universities.

Reasons for the over-great respect for Aristotle's logic were at first his system's being the only one of which one had records making possible detailed knowledge, together with the authority of the Church of Rome which had taken up, with some adjustments, much of Aristotle's teaching in philosophy; and later the crushing effect of higher education which was controlled by those who had given their years to the learning of the languages, history and writings of the Greeks and Romans and so naturally put the highest value on such knowledge, a value much higher than they were ready to give to new sciences and to new, natural knowledge.

An example of a syllogistic implication having validity is 'if all animal is going to die and all man is animal, then necessarily all man is going to die'. This implication is of the form 'if P is truly said of all M and M is truly said of all S, then necessarily P is truly said of all S'. (The reason for saying 'man' and 'animal' and not 'men' and 'animals' will be made clear later, in 6.2.)

2.6. The use of variables by Aristotle. It was in such forms of syllogistic implications that Aristotle was interested and rightly for Syllogistic because the business of this is with the laws of syllogistic implications, and if these laws are to be general, it is necessary for them to be put forward in relation to the structures and not in relation to material examples of the implications. One has to give Aristotle great credit for being fully conscious of this and for seeing that the way to general laws is by the use of *variables*, that is letters which are signs for every and any thing whatever in a certain range of things: a range of qualities, substances, relations, numbers or of any other sort or form of existence. The S, P and M higher up are examples of the variables used in Syllogistic. In Aristotle's theory the range of such variables is, at the level of language, the range of all possible general names and, parallel to this, the variables are, at the natural level, the representatives of any qualities and of any sort of substances, like star and metal and backboned animal.

8

Aristotle's Syllogistic

If one keeps in mind that the Greeks were very uncertain about and very far from letting variables take the place of numbers or number words in algebra, which was why they made little headway in that branch of mathematics though they were good at geometry and so why the invention of a geometry based on algebra was not made till number variables had become common and were no longer strange, in the time of Fermat and Descartes, then there will be less danger of Aristotle's invention of variables for use in Syllogistic being overlooked or undervalued. Because of this idea of his, logic was sent off from the very start on the right lines. But not before Boole, 2200 years later, who saw in the old Syllogistic the seeds of an algebra of logic, was the important step taken of changing Syllogistic into a mathematics of certain reasoning. Only when logic was married to mathematics did it become fertile. On its becoming the mother of a good number of ideas and teachings of profit, logic's position as the poor relation of philosophy came to an end, the change going so far as to make the new logic widely looked at and respected as the head of the family. In fact, it has been the desire of some persons to see the rest of the family dead. This is true of the supporters of the Vienna Circle that was flowering in the 1920's and 1930's: it was their belief that logic, in the wider sense that takes into account, in addition to Mathematical Logic, the logic of chance and the logic of the natural sciences—logics which have again, like Mathematical Logic, deep connections with mathematics—is the one part of philosophy that is of serious interest and value, because it is the only one in which true knowledge is possible; in the other parts man's thought is wasted on systems of fictions caused by the tricks of language and on talk which, though put forward as offering a key to the secrets of all that there is, is without good sense, by reason of its being framed in such a way that its powers of opening the locks of existence are unable to be publicly tested by experience.

2.7. Aristotle's errors. What are the points of Aristotle's Syllogistic where he has gone wrong or where one might do better? These points may be put into two groups, one group having those which come inside Syllogistic itself and the other having those points which are outside Syllogistic.

Some points in the first group are these. In the old logic only two general statements as conditions and, in reasonings having validity, not more than three general names may be used in implications. But there is no good reason for so limiting implications while there is a good one for not so doing, this being that the business of logic is with every sort of implication having validity and there are implications having validity which have more than two general statements as conditions and make use of more than three general names; an example is 'if all animal with wings is backboned, no self-moving thing is rooted in the earth and all that is backboned is self-moving, then necessarily no animal with wings is rooted in the earth'. Again, Aristotle's logic is not a well worked out system like the systems of mathematics; in these the different statements are or are made to be dependent on one another: one takes a selection of some of them as starting-points from which the rest have to be got by deductions. Though, it is true, Aristotle's Syllogistic is different only in degree from for example Euclid's system of geometry, still the degree between them is enough for it not to be unkind to say that Aristotle's theory is nearer to a simple grouping than to an ordering and organization of his material. In addition, its use of signs other than words not being taken far, the old logic was not an algebra and so it was not much help in the art of quickly getting the right answers to questions about reasoning and of making fully clear that the answers were right because of the processes of deduction by which they were produced, processes based on a body of rules fixed at the start.

There are two chief points in the second group about which we will say something, again without going into details. Aristotle and his school of logic made no attempt to give an account of the theory of implications having structures whose validity is dependent on the force of words like 'not', 'and', 'or' and 'only if' which are used for forming more complex statements from given ones; an example of such a structure is 'if s_1 is false but s_1 is true or s_2 is true, then necessarily s_2 is true', and another example is 'if s_1 only if s_2 but s_2 is false, then necessarily s_1 is false'. This branch of logic may be named the logic of statement connections. It was first given attention by the Megarians and Stoics, working quite separately from Aristotle; see 2.8. Secondly, the old logic did not take into account implications such as 'if A is greater than B and B is greater than C, then necessarily A is greater than C', 'if A is B, then necessarily A of X is B of X' ('if song is music, then necessarily the song of a bird is the music of a bird') and 'if A is B, then necessarily X of A is X of B' ('if Christ is the Highest Being, then necessarily the mother of Christ is the mother of the Highest Being', an example given by Leibniz), which are dependent on words or ideas for relations between things. The expansion of the old logic so that at least some part of the logic of relations might be covered was first undertaken by De Morgan in the 1840's; in his view the logic of relations was more important

11

because wider than Syllogistic, this being nothing but a small part of that logic.

2.8. The Megarian and Stoic logic of statement connections.

We will here say something about the logic of the Megarians and Stoics; that logic was first formed at about the same time as Aristotle's, and it will be best if this present history of the development of logic keeps as far as possible to the order in time in which the different schools came into existence.

The Megarian work on logic was done between, roughly, 400 and 275 and the Stoic between, roughly, 300 and 200 before Christ. There were two chief ways in which the Megarian-Stoic theory was different from Aristotle's. Firstly, it was interested in forms of reasoning having the structure of an argument and not the structure of an implication. Secondly and more importantly, it was a logic of statement connections and not one of general and suchlike statements.

The sorts of reasoning to which the Megarians and Stoics gave attention were these: 'if s_1 then s_2, and s_1; so s_2'; 'if s_1 then s_2, and not s_2; so not s_1'; 'not the two of s_1 and s_2, and s_1; so not s_2'; 's_1 or s_2, and s_1; so not s_2'; and 's_1 or s_2, and not s_1; so s_2'.

Some of these and other forms have an 'if-then' statement as one of the bases of the argument. Because one may rightly put forward as true what comes at the end of an argument only when all the statements on which it is based are true, it is necessary for an answer to be given to the question: when is an 'if-then' statement true? This question was much talked about in the Megarian-Stoic school, so much so that someone (Callimachus) said: 'Even the birds on the rooftops are having discussions on the question.' A number of different answers to it were offered; we will give three

12

of them. Philo said that all 'if-then' statements are true but for those which have a true 'if' part and a false 'then' part, these being false; so 'if $2+2=5$, then it is day' is true at all times, and when it is daylight 'if it is night, then it is day' is true while 'if it is day, then it is night' is false. Philo's sort of 'if-then' relation has been that normally used in Mathematical Logic; it has been named by Russell 'material implication'. Again, Diodorus said that an 'if-then' statement is true if and only if it is one which at no time has a true 'if' part and a false 'then' part; so 'if it is night, then it is day' is not a true statement even if made in daylight because there are times—at night—when 'it is night' is true and 'it is day' is false. Because, in Diodorus's opinion, an 'if-then' statement is true if and only if it is at all times a true material implication, his sort of 'if-then' relation may be named 'all-time material implication'. And thirdly, there were some who had the belief that it is right to put forward an 'if-then' statement as true only when there is some connection between what its two parts are about; they said that 'if s_1, then s_2' may be judged to be true when and only when s_2 is necessarily true if s_1 is true, or, putting it another way, when it is not possible that s_1 is true and s_2 is false. This sort of 'if-then' relation is what was named simply 'implication' in 2.3.

THE IDEA OF A COMPLETE, AUTOMATIC LANGUAGE FOR REASONING

3.1. Lull's 'Ars Magna'. The first step in the direction of a complete and automatic language for reasoning was taken by Ramon Lull (1235–1315) in about 1270, in his book *Ars Magna* ('The Great Art'). Lull's belief was that all knowledge in the sciences is a joining together of a number of root ideas: knowledge is a complex of simples. There were only 54 of these root ideas, about a third of them having to do with the field of religion or of theories about right and wrong behaviour. The joining together of groups of these ideas is 'the great art' by which *scientia generalis* ('the substance of science') is to be effected. Lull did not get much further than working out the number of ways in which complexes of these ideas might be formed. He gave no rules, or only foolish ones, for judging the value as knowledge of the different possible complexes. It seems to have been his opinion that no knowledge in the sciences has any need of sense-experience as a guide and a support, as if the discovery and the testing of the discovery of what is under the sea might be made without stepping from the land: the fishing-boats of knowledge may be kept safely and with profit in harbour and do not have to be sent sailing with their nets over the deep waters of possible experience. In having this opinion, Lull was representative of much of the thought of his time.

3.2. Dalgarno's and Wilkins's languages. After Lull suggestions for a general language for a general science were not uncommon in philosophy. But the development of such a language was not made till the 1660's, when the *Ars Signorum* ('The Art of Signs') of George Dalgarno (1626–1687) and the *Essay towards a Real Character and a Philosophical Language* of John Wilkins (1614–1672) were printed. The ideas at the base of Wilkins's *Essay* were, to some degree, copied from Dalgarno. However, it is not by chance or without good reason that Wilkins's book has been better kept in memory than Dalgarno's. The working out in detail of the ideas for a new consciously made language is more valued than an outline of them and the *Essay*, the last and separate part of which (by William Lloyd) gives a full list of the English words of the day and their senses in Wilkins's language, is on a much greater scale than Dalgarno's. Things were not so hard for Wilkins, who at one time or another was a Head of Colleges at Oxford and Cambridge—the only man ever to have been so at the two places—and who had a high position in the Church of England; he had the backing of the newly formed Royal Society, of which he was one of the Secretaries, and his book was printed by the order and the necessary money came out of the pocket of the Society. Dalgarno, on the other hand, was a private school teacher.

Wilkins's purpose was, he says, to give a regular list and account of all those things and ideas on which marks or names might be put in agreement with their natural properties, it being the end and design of the different branches of science that all things and ideas

15

be placed in such a frame as may make clear their natural order, their ways of being dependent on one another and their relations. (It is easy to see here the effect of Bacon's teaching on the roads and limits to knowledge in the sciences: science is the discovery and ordering of natural groups of facts, not the building of delicate theories of them.) Wilkins goes on to say that an art of natural rules for forming statements and questions is needed from which one will get 'such helps and instruments' as are necessary for the framing of the simpler ideas into smooth and unbroken talk or writing. The smaller and simpler this body of natural rules is, the better. Wilkins had three points in mind: the rules of his language were to be 'natural'; the different words were to be so formed as to be dependent on and in those relations to one another as would make them parallel to the things and ideas of which they were 'marks or notes'; and the forms of these names were to be ordered, by their letters and sounds being like or opposite to other letters and sounds, so that the names would in some way be in agreement with the things they were signs of. In addition, Wilkins, like Dalgarno before him, had the opinion that a system of shorthand writing, in which the shorthand outlines are dependent only on the sounds of the words, was needed for a specially designed language such as his and no small part of his book is given to the development of this side of his language.

The great trouble Wilkins took to make his language a living instrument of science (and international trade and religion) was completely wasted—but for its helping to keep before men's minds the thought that it is possible for a language better than the everyday ones to be formed by conscious art. The offspring of Wilkins's brain was dead at birth. One reason for this was that because of discoveries in science and the changes in common knowledge caused by these it came to be seen that his accounts of things in relation to 40 chief headings and under these headings an attempted grouping of all other things by divisions of higher and lower levels, in this way giving an account of everything by what it has in common with certain other things and how it is different from them, as in taxonomy—were false or without support; and, equally serious, it was seen that the new ideas and sorts of facts that were the outcome of the new knowledge were not covered by Wilkins at all, while they put in the shade much of what was covered, for example about the Highest Being and other such ideas that were a necessary part of the beliefs of a man having authority in the Church. Another reason why Wilkins's system had almost no effect is that the power of language, in the sciences, of giving and ordering natural knowledge was very much overvalued by him. On condition that a language is able to be used for making as sharp and as detailed recordings of observations as is desired and is able to undergo expansion, letting new ranges of fact and of thought be talked about without unnecessary trouble being caused by the language apparatus, then it is not generally important for the natural sciences if the language is not as good as possible in other ways, for example in having forms that are not completely regular. On the other hand, languages, such as those of mathematics, having short and simple signs and for which there are rules for building statements which put forward two groups of numbers or of other things as being equal or being unequal and for going from some given statements of this sort to other ones, by processes of deduction, do have great powers of increasing and ordering knowledge. But this knowledge is of possible

17

relations between things, not of facts themselves. Wilkins did not see that if a designed language was to be of use in the way he was picturing to himself, it would have to be a sort of arithmetic or algebra. However, he was not expert enough in mathematics to be able to make the necessary developments for this new purpose.

3.3. Descartes' idea of a general language. It seems that Descartes (1596–1650) was the first person to have the idea of a general language—'une langue universelle' are his words for it—as a sort of arithmetic. In a letter to Mersenne of 20 November 1629 he said that the invention of a language is possible in which 'an order is formed between all the thoughts which might come into our minds, in the same way as there is a natural order between numbers; and as learning how all the numbers, without end, are named may become complete in no more than one day . . . all these names being different words, so the same may be done in connection with all the other words necessary for talking about all the other things coming in our minds. . . . The invention of this language is dependent on the true philosophy, because but for this it is not possible for all the thoughts of men to be listed and put in order or even to be made separate from one another so that they become clear and simple, which is in my opinion the great secret of getting well-based know-ledge. And if someone was to give a good account of which are the simple ideas that are in men's minds of which all their thoughts are made up . . . there would be almost no chance of going wrong.' No doubt wisely Descartes made no attempt to give a list of all our simple thoughts and to put them in order so that an arithmetic of reasoning might be formed that would

Language for Reasoning

let one get complete and certain knowledge of whatever is true.

3.4. Leibniz's idea of a mathematics of thought.

A little later Leibniz (1646–1716) had designs for a new and general language that were not unlike Descartes', but he took them some stages further. From as early as his *De Arte Combinatoria* ('On the Art of Complex Forms'), printed in 1666, Leibniz made suggestions for a mathematics of ideas: as every number greater than 1 is in the range 2, 3, 5, 7, 11, 13, . . . of numbers into which division is possible only by 1 and itself, or is a number that may be produced from numbers in that range (12=2×2×3, 35=5×7), so all ideas are simple ones or, if they are not simple, are complex ones that may be formed by putting together some of the simples. The simple ideas are noted by numbers in the range 2, 3, 5, 7, 11, 13, . . . while the complex ideas are noted by numbers in the range 6 (=2×3), 10 (=2×5), 14 (=2×7), 15 (=3×5), 21 (=3×7), 22 (=2×11), 26 (=2×13), 30 (=2×3×5), . . . , the number for a complex idea being dependent on the numbers representative of the simple ideas from which it is formed. Talking at a later time of his early work, Leibniz said he was forced by a sort of inner driving to the view that 'the invention of an ABC of man's thoughts was needed, and by putting together the letters of this ABC and by taking to bits the words made up from them, we would have an instrument for the discovery and testing of everything'.

Leibniz was ruled by a belief in mathematics as the key-science; it seemed to him that the operations of mathematics are able to undergo a wide enough expansion for decisions on all questions to come under its control 'in so far as such decisions are possible by reasoning from the given facts. Because though some experiences are ever needed as a base for reasoning, when these experiences have been given we would get from them everything that any person would ever be able to get from them, and we would even make the discovery of what more experience is necessary if our minds are to become free from the rest of our doubts. . . . If we had a body of signs that were right for the purpose of our talking about all our ideas as clearly and in as true and as detailed a way as numbers are talked about in Arithmetic or lines are talked about in the Geometry of Analysis, we would be able to do for every question, in so far as it is under the control of reasoning, all that one is able to do in Arithmetic and Geometry. All work in the sciences which is dependent on reasoning would be done by the changing and exchanging of signs and by a sort of Algebra; an effect of this is that the discovery of facts of great interest and attraction would become quite straightforward. It would not be necessary for our heads to be broken in hard work as much as they are now and we would certainly be able to get all the knowledge possible from the material given. In addition, we would have everyone in agreement about whatever would have been worked out, because it would be simple to have the working gone into by doing it again or by attempting tests like that of "putting out nines" in Arithmetic. And if anyone had doubts of one of my statements I would say to him: let us do the question by using numbers, and in this way, taking pen and ink, we would quickly come to an answer.'

3.5. Leibniz's errors. The idea of what we have been naming a complete and automatic language for reasoning is made very clear by these words of Leibniz. But though he frequently gave attention to the idea of such

a language he did little with it. He was at the same dis-
tance from having a new language system at the end as
he had been in 1666. His designs got no further than
being designs—and rough ones. Leibniz went wrong in
two ways. Firstly, he was unwise not to have made much
narrower the field which was to be covered by the
machine-like operations of his desired invention.
There is no hope of covering from the very start every
part of what might be reasoned about; the building of
any such apparatus as was pictured by Leibniz, as of
any theory in science, comes only with a slow process of
growth, not jumping suddenly into a condition of full
development at birth like Athena out of the head of
Zeus. It is necessary to go one step at a time, starting
with what is simplest; this is the best way of getting
knowledge that is of value in itself and of value as a base
for making wider the range of facts or thoughts taken
into account. Leibniz would have done much better by
limiting himself to an attempted forming of a language
for no more than one or two branches of the mathematics of his day. Desiring to have a system that was a
guide to everything, he was unable to put together a
system that was a guide to anything. At the back of this
desire was the belief, by which Descartes earlier had
been gripped, that to a person like himself who had
made such new and important discoveries in mathematics the invention of an instrument of logic having
the power to give the answers, without error and without waste, to all that is a question for man's reason, was
a work of little trouble and of little time. This belief was
one side of the thought then current: in living memory a
great increase of knowledge had taken place and the
mind had become much more free and open, had
become forward-looking and full of hope, sharply conscious of its powers.

21

The Idea of a Complete, Automatic

Secondly, Leibniz went wrong by desiring to have two things at the same time: on the one hand, to have a system of the forms and structure of reasoning as such —a system of logic, simply—and, on the other hand, to have a system in which all that might be the substance of some argument would be ordered—a system of physics and philosophy. The reward would have been greater if he had made the decision to keep to the logic, which is so much more straightforward and which would have given him the chance of turning the metal of the old logic into the gold of a new and better one, for example one whose laws, unlike those of Syllogistic, are enough and have the right properties so that the validity of arguments in at least some of the theories of mathematics may be tested and judged. Or, if Leibniz had the feeling that the material in addition to the form of reasoning had to be a part of his language system, he might have done well—much better than he did— by letting himself be interested in the bricks of only mathematics, having as his purpose an organization of mathematics based on its simplest ideas and starting-points. A road to the full development of such an organization was not seen till about 200 years later: by Frege, Peano, Whitehead and Russell, in the years between 1875 and 1910.

It would not be right to get the idea that Leibniz's work was of no value at all because his controlling thought of a new mathematics-like language came to nothing. Having that thought, it was natural that his mind was transported in the direction of logic itself, and there he made a number of important discoveries, some of them in Syllogistic but most of them in fields of logic that were new or had been given little attention before. However, these discoveries were not taken seriously till much later because the separate bits were not united in a general theory and so, not being complete enough, they were not put in print or offered to the public in Leibniz's time or even till long after; by then they were common knowledge, not as his teachings but as the teachings of theories worked out in the years after 1840, theories whose start and development were in no way dependent on Leibniz's work on logic.

4

CHANGES IN ALGEBRA AND GEOMETRY, 1825—1900

4.1. Peacock's algebra. Between 1825 and 1900 algebra and geometry underwent great changes, changes producing by 1900 a completely different general outlook in the philosophy of mathematics from that which had been common before. The changed forms and purposes of algebra and geometry had a strong effect on all the parts and at every stage in the growth of Mathematical Logic.

An *equation* is a statement that two things, normally two groups of numbers or of signs representative of numbers, are equal. Till 1825 or a little after that, algebra was nothing but the theory of equations in which letters were used for numbers and such signs as $+, \div, -$ and $\times$ were used for the four operations of addition and division and their opposites, the equations being those of a sort that the reader was probably learning about in high school, for example $3x-2y=11$

and $ax^2+bx+c=0$. The business of the theory was to
get a knowledge of how such equations may be worked
out to give number values which make them true and
to get a knowledge of the conditions controlling the
existence of and relations between the number values.
The four operations of addition and so on were done,
as they are still done by schoolboys, more or less un-
consciously by making whatever moves seemed right
and natural, the rules supporting these moves resting
in the dark. There was no thought that a statement of
the rules was necessary or might be a help in the
development of algebra. It is strange that though
geometry had early in its history been turned into a
structure of fixed starting-points and of reasonings from
these, the position of algebra, like arithmetic, was
different. The need of forming a table of and of being
consciously guided by the laws of algebra which say
what are the properties of the operations in algebra was
seen only by Peacock (1791–1858). In his book *A
Treatise on Algebra* (1830; second writing, in two parts,
1842–1845) the idea was put forward that algebra is,
rightly viewed, a science of deductions like geometry.
(At about the same time—1837—the first account of
Newton's mechanics as a science of deductions, copying
Euclid's *Elements*, was given by Whewell, a friend of
Peacock, in his *Mechanical Euclid*.) Peacock had two
chief points which are to be noted here. Firstly, all the
processes of algebra have to be based on a complete
statement of the body of laws which are about the
operations used in those processes, no property of an
operation being used if the operation's having that
property has not been forced into the open and been
taken as true in a law from the start or has not been got
by deduction from the first laws. Secondly, the signs
for the operations have, at any rate for the purpose of

making deductions, no senses other than and more than those that are given them by the laws. No properties of for example addition may be used if they are not among the listed properties of addition. The lines of deduction are limited by, but may be stretched so that as wide a field as is desired is covered if they are kept in agreement with, the forms of the laws. An effect of this is that addition is any operation at all with the properties put forward in the laws of algebra having in them the sign + (or whatever the sign for addition may be). From this point of view came the new sort of algebra named by Peacock Symbolical Algebra and later given the name Abstract Algebra. Boole who was chiefly responsible for the birth of Mathematical Logic was deeply acted on by Peacock's ideas which made possible the building up of new algebras, that is systems with laws not unlike those of school algebra but which one has in mind as being about things other than common numbers, and especially the building up of the algebra of logic.

Examples of the laws of algebra are the law that in an addition the outcome is the same even if the order of the variables is changed: $a+b=b+a$; the law that in additions the outcome is the same even if the grouping of the variables is changed: $(a+b)+c=a+(b+c)$; and the law that when a is not equal to 0, then $b=c$ if $a \times b = a \times c$.

4.2. Equations of the fifth degree. Another event causing the movement in the direction of Abstract Algebra and away from the limits of the old theory of equations was the surprising discovery by Abel and others that number values for equations of degree 5, such as $3x^5 - 2x^4 + 7x^3 - 19x^2 - 4x - 1 = 0$, are generally unable to be got by the arithmetic-like processes—

specially the taking of square and higher roots—which make possible the automatic getting of number values for equations of degree 1, 2, 3 or 4. The belief had become fixed that answers to questions about the number values for equations of higher degree than the fourth would be (and anyhcw that it would not be possible to make certain that they may not be) got in the same simple way as for those equations of no higher than the fourth degree. Abel's discovery was seriously damaging to the belief, having a wide distribution among workers in science and philosophy, that at least the heart of future knowledge will be in agreement with the expert knowledge of their time. This belief had to take blow after blow, not only in mathematics but in the natural sciences, through the years after this discovery in algebra, some of the blows coming from Mathematical Logic.

Galois, in attempting to give an account of the reasons why the facts were as Abel had said them to be, had the idea of a group (in a special sense of the word 'group' used in mathematics) and saw some of the important properties of groups. Guided by his work a separate and fertile part of algebra—Group Theory—came quickly into being as the first new branch of Abstract Algebra. In this way the changes in algebra taking place in England came together in harmony with those taking place in other countries of Europe.

4.3. Vector and matrix algebra. To those who were pained by the discoveries of Abel and Galois, worse was to come. They were given more, violent shocks by the systems of Vector Algebra and Matrix Algebra worked out not long after the death of Galois in 1832. These systems were company for Group Theory as new branches of Abstract Algebra.

Changes in Algebra and Geometry, 1825–1900

In mechanics and electromagnetics an important part is played by the idea of properties, of which force is one, that not only have a certain size in relation to some unit on a given scale but have a direction in relation to some framework of positions. The algebra of measurable properties acting in a direction, that is Vector Algebra, was formed by W. R. Hamilton and Grassmann in the 1840's. In their theory $\mathbf{a} \times \mathbf{b} = \mathbf{b} \times \mathbf{a}$ is not generally true though the operation $\times$ is in some other ways like the operation $\times$ of number algebra; for example it is true in Vector Algebra as in school algebra that $\mathbf{a} \times (\mathbf{b} + \mathbf{c}) = (\mathbf{a} \times \mathbf{b}) + (\mathbf{a} \times \mathbf{c})$. (Signs for direction-properties and for operations on them are normally put in blacker print.)

The 1850's saw the invention of Matrix Algebra by Cayley. This branch of mathematics is the theory of number tables having m lines of numbers across and n lines of numbers down, so that a table has $m \times n$ numbers in it. On condition that the number n of lines across in the table B is the same as the number of lines down in the table A, an operation $\times$ is possible on A and B, that is another table $A \times B$ is able to be produced from A and B. But as in Vector Algebra the 'law' $A \times B = B \times A$ is not generally true. To take a simple example: let $A = (1 \quad 2)$ and $B = \begin{pmatrix} 3 \\ 4 \end{pmatrix}$; then by the sense given to $\times$ in connection with number tables $A \times B = (1 \quad 2) \begin{pmatrix} 3 \\ 4 \end{pmatrix} = (1.3 + 2.4) = (11)$, while on the other hand $B \times A = \begin{pmatrix} 3 \\ 4 \end{pmatrix} (1 \quad 2) = \begin{pmatrix} 3 & 6 \\ 4 & 8 \end{pmatrix}$; it is clear that these two number tables got by working out $A \times B$ and $B \times A$ are different, two tables being equal—the same—if and only if they have the same number of lines across, the same number of lines down and what is given in one table at the meeting point of any line across and any line down is the very same as what is given in the other table at that point. It was at least in part because of the birth of new systems in which old laws like $a \times b = b \times a$, which had for so long been taken as true and nowhere open to doubt, were not true, that one became conscious of the need of making clear what were the starting-points and the rules used in the different fields of algebra. One saw that what was true was dependent on the field and that there were more fields than the old and common one. This change in algebra, whose full growth took 100 years (1830–1930), was of great weight in the development of almost every other part of mathematics. To give only one example of the way in which Mathematical Logic was touched by it: because of the new current of ideas Schroeder and Whitehead in the 1890's were caused to give an account of the algebra of logic in the form of a science of deductions using to no small degree the signs and processes of algebra and of its latest directions in Abstract Algebra. These workers, in their turn, were copied by others, and it became the normal thing in Mathematical Logic to take note of the latest theories in algebra, by which it is offered new food for thought that keeps it a living branch of philosophy and science.

4.4. Geometries of Euclid and Lobachevsky. For the present purpose the history of geometry has three chief turning-points. The first was the making of geometry, by Euclid in his *Elements* (about 300 years before Christ), into a system of reasonings from a fixed base, taken to be true without the support of reasoning. By his doing this, geometry was changed and kept from being nothing but a mass of separate statements and was turned into a united body of knowledge, all the parts of which were dependent and in connection with one another. Because of Euclid's work geometry was looked at down the years as the most complete branch of mathematics and its deduction form was looked at as offering the best way to true knowledge and, outside religion, the highest knowledge.

The second turning-point was the invention by Lobachevsky in 1825 of a geometry that is different in its roots from Euclid's. Their systems are unlike in this: in a plane, through any point not on a given straight line there is in Euclid's system one and only one while there is in Lobachevsky's more than one straight line parallel to the given line. Other ways in which they are different are outcomes of this; for example the law of Pythagoras is true and may be got by deduction in Euclid's but is false and may not be got by deduction in Lobachevsky's geometry. The fact that one was able to have a system other than Euclid's came as a deep shock, producing in a quite short time a view of what mathematics is about, what its purpose is and what its relations to physical space and things in that space are which was completely unlike the normal, old view in the philosophy of mathematics. When it was seen that one may have two geometries where what is true in one is false in the other, the old opinions such as Plato's, the Schoolmen's, Spinoza's, Leibniz's and Kant's by which mathematics is a science giving completely certain knowledge of existence—natural or higher-than-natural, and if not of things-in-themselves, then of things as they seem to be—were not able to be put forward with reason on their side.

The effect of the changed outlook on mathematics and specially geometry was to make theories in mathematics viewed not as in themselves natural knowledge but as in themselves works of art, dependent for their value on their attractions for the mind, though they may sometimes be judged to have another value in addition: that of being a help in the natural sciences. Mathematics is the attempt to get interesting ideas about numbers and spaces and suchlike things, to make interesting discoveries about their properties and relations, and to put the teachings supported by these discoveries into systems of reasoning. Mathematics is the free invention of the mind, limited only by possible uses in natural science that may be desired and by rules of logic that one is willing or is forced, as a person of good sense, to be guided by.

The third turning-point in the history of geometry is best talked about in connection with what comes under the fourth and last of the headings given in 2.1, so we will now go on to that.

5

CONSISTENCY AND METAMATHEMATICS

5.1. Mathematics and systems of deductions: consistency. The idea of the parts of mathematics as systems of deductions goes back to Euclid's work in geometry, geometry being the first part of mathematics that was made into a system of this sort. What a system of deductions is has been outlined clearly enough to the reader for us not to have to go into more detail

Consistency and Metamathematics

about it at this stage, though more will be said later. What we will have to say something about here is another side or level of a system of deductions, the idea of which has been very important for Mathematical Logic.

Some attention in relation to geometry had been given before the 1890's to the question: Is it possible to get at the end of one chain of deduction in a system a statement *s*, which becomes by this fact a *theorem*—a true statement—of the system and to get at the end of another such chain the theorem not-*s* which says that *s* is, after all, false? A system of deductions is said to have the property of *consistency* if and only if it does not have two opposite theorems *s* and not-*s*. The amount of attention that had been given to the question of consistency had been small because the question had not been taken very seriously, the belief being general in mathematics that any expert was able quickly to see the answer to the question of a system's consistency without his having or needing a supporting argument. But in the 1890's the question was viewed as interesting and important by that Italian school of mathematics that was headed by Peano. From the point of view of logic there is no doubt that the question is an important one because any statement whatever about the field under discussion becomes a theorem in a system that does not have the property of consistency, the statements being true ones or being false ones about the field, so that what is true would not be able to be kept separate from what is false, while the very purpose of a system is to have true and only true statements about the field as theorems. Because the question of consistency is important, how the answer is made and supported is important. Peano's opinion was that reasoning and clear public tests are the right instruments for making decisions about consistency. (See end of 9.4.)

He had the same opinion about another question that might be put about a system of deductions. Giving the name *axiom* to a statement that is taken to be true, without the support of reasoning, at the base of such a system, this question is: Are any of the axioms of the system dependent on the others in the sense that it is possible to get an axiom as a theorem at the end of a deduction in which the axiom itself is not used?

Some thought had been given to this question and the question of consistency from quite early times in connection with attempts to get Euclid's axiom of parallels as a step in a deduction not resting on that axiom, these attempts being made to let one see that that axiom is dependent on the other axioms: which it is not. However, only in the 1890's were these questions regularly put, and put about all systems of deductions in mathematics, and were ways designed for answering them by arguments. Peano and his school were chiefly responsible for this, though from late in the 1890's the work, in the same direction, that was done by Hilbert in Germany had a greater effect on later developments.

5.2. Metamathematics. From this work came a new branch of Mathematical Logic: metamathematics. The business of metamathematics is with what may be truly said about some system (or group of systems) of deductions, for example that it has the property of consistency or that no axiom is dependent on the others. If S is a system whose field under discussion is F, then the purpose of S is the building of a framework of supportable theorems about F, the theorems saying what true relations there are between the things in F. If S is an arithmetic, then a theorem in S such as '7+5=12' is about numbers. On the other hand, ' "7+5=12" is a theorem in *S*' is not about numbers; it is a statement about a statement about numbers. '7+5=12' is a theorem in *S*; ' "7+5=12" is a theorem in *S*' is not a theorem in *S* but a theorem in the metamathematics *S'* of *S*. It is necessary for the reader to see what this delicate point is that makes *S* and *S'* different from one another. Every system *S* has its metamathematics *S'*, sometimes named its 'logic', which is formed of the true statements, and the arguments for them, that may be made about *S*, while *S* itself is formed of the true statements, and the arguments for them, that may be made about *F*; for example, if *S* is an arithmetic of the common sort, then *F* is the body of all the numbers such as 34, 1/2, − 6/79. *S* is normally a part of mathematics, though systems of deductions outside mathematics are possible; *S'* is a part of logic: the logic of that part *S* of mathematics. The birth of Mathematical Logic was in the 1840's. Most of its more important and more interesting developments after its first 50 years of growth have been at the level of metamathematics.

6

BOOLE'S ALGEBRA OF LOGIC

6.1. Boole and the older logics of Aristotle and of Hamilton-De Morgan. Now that the four currents of thought named in 2.1 have been talked about, we may have a look at that stretch of Mathematical Logic which their motion and driving-force have so far produced;

as one might say, the look will be given from the foot-way at the side of the river and not from the point of view of one having a swim underwater, because clearly we are not able to go deeply into Mathematical Logic here.

What was the start of Mathematical Logic? The shortest and simplest answer is George Boole's *Mathematical Analysis of Logic* (full name: *The Mathematical Analysis of Logic, being an essay towards a calculus of deductive reasoning*) of 1847. There is nothing completely new under the sun. Every birth is the outcome of earlier events. So Boole's little book of 82 pages was only marking a stage in an unbroken line of thought from the past. However, though it certainly had connections with what some others had done, it was still different enough from the rest for it rightly to be seen as starting a quite new theory—the theory of Mathematical Logic—and not as being another step in an old one.

The earlier teachings in logic of which Boole had a knowledge and which had an effect on him were, on the one hand, those of the old logic and, on the other hand, those of Sir William Hamilton (1788–1856) and De Morgan (1806–1871) on the theory that was based on the changing of the four, *A*, *E*, *I* and *O* forms into a greater number of forms in which the amount of the predicate is given; for example, Hamilton has two *A* forms, one being 'all *S* is all *P*' and the other being 'all *S* is some *P*'. For later developments two points were important in the theory of Hamilton and De Morgan (the two of whom had a bitter fight, Hamilton protesting that his ideas had been taken from him without his being given credit for them, and De Morgan answering that his account was not at all dependent on anything that Hamilton had ever said). One point was the weight placed by it on the amounts of predicates in statements, the old logic noting only the amounts of subjects. The value of this was that every statement of the subject-predicate form was able to be turned into an equation or into a statement that an equation was false. By placing this weight on the idea of an equation, logic was moved near to algebra. Secondly, what had been named by Aristotle the two 'ends', S and P, of a statement had generally been viewed as signs of qualities. In the new theory of Hamilton and De Morgan S and P were changed into being signs of the things themselves that have the qualities. An example of an A statement in the old logic is 'all leaf is green', this having the sense that the quality of green is a 'part' of the quality of leaf. From the point of view of the new theory this statement had the sense of saying that whatever thing is a leaf has the property of being green: 'all leaves are green things'; or, putting it another way, 'the group of all the things that are leaves is a "part" of all the things that are green'. Putting the change shortly, it was a change from 'all S is P' to 'all S's are P's'. Though the reading of 'all S is P' as 'all S's are P's' had sometimes taken place before Hamilton and De Morgan (for example in Whately's *Logic*) the reasons for so doing, when there were any conscious reasons, were not reasons of logic. And the greater minds, such as Leibniz's, had kept to the S and P form, not making use of the S's and P's form. It was because they kept to the S and P form that they were stopped from building a better logic. The better logic they had hopes of building was to be a logic of mathematics, even if for more than of mathematics, and it was to be one based on mathematics. However, reasoning in mathematics is generally not about qualities as such but about the things that have the qualities. For

35

example, geometry is not a theory of the qualities 'being a point', 'being a line' and 'being a plane'; it is a theory of the things that are points, lines and planes.

The name used in logic and mathematics for a group of all the things that have a certain simple or complex property is *class* and the things that have the property are said to be the *elements* of the class. The ideas of class and class elements are root ideas in all present-day mathematics. The outcome of the Hamilton–De Morgan theory was to make possible a view of logic as being, at least in one of its branches, an algebra of classes. Boole was the first man to have this view clearly.

6.2. Boole the man. Boole (1815–1864) was, till he went across the sea to Ireland as the first Professor of Mathematics at Queen's College, Cork, when this was started in 1849, an English private schoolteacher, whose schooling had come to an end when he was only a boy and who had at no time any sort of higher education. His father was in trade on a very small scale, as a shoemaker, and the family had little money. The older Boole was a man of regular behaviour, living with care, and having a love of learning and a deep interest in the sciences. These qualities were handed down to his son, who got from his father, in addition, a knowledge of the simpler parts of mathematics. George Boole became a teacher for his living when he was only 16 years old. By teaching himself from books in his free time from teaching others, he quickly became a person of wide learning, specially expert in mathematics. When he was 24 he was writing papers on new mathematics for the *Cambridge Mathematical Journal* and five years later the Royal Society gave him a special mark of reward for the very first paper he sent it. He is a good example of what might be done by self-help.

Boole's Algebra of Logic

6.3. Boole's 'Mathematical Analysis of Logic'.
Boole's first of four books was *The Mathematical Analysis of Logic*, whose writing took only some weeks in the spring of 1847, when he was 31. This is a work offering a logic based on mathematics, chiefly algebra; its root ideas are those of class and class elements and of operations of selection of elements from classes; and its theory about these is formed by the use of equations. What he does is to give an account of the old logic as an algebra. It is pointed out how *A, E, I* and *O* statements may be put in the form of simple equations; how the necessary outcomes of any one of these may be got by algebra from its equation, for example how 'no *P*'s are *S*'s' may be got from 'no *S*'s are *P*'s'; how the validity of a syllogistic implication may be tested by turning the group of statements in the implication into a system of simple equations and seeing if the outcome's equation may be got by algebra from the equations of the conditions; and how, if certain statements are given as conditions of a syllogistic implication, but no outcome is given, it is possible to get by algebra a necessary outcome of them from their equations—if they have any necessary outcome.

But, and this was a great step forward, Boole gave in addition an account of the logic of statement connections (2.7) as an algebra; unlike Leibniz, Hamilton and De Morgan, he saw that this part of logic was important, and where others had been completely at a loss he was able to give a good theory of it. It is interesting that his theory was the very same in form as that of the algebra of classes. Boole was the first man to give a united theory of logic.

We have said that Boole's algebra of statement connections is the very same in form as his algebra of classes. This is true of the body of the book. But after

37

this had been printed an addition of some pages to make clear a number of points in it was put in, and on the very last page Boole made a note of the fact that there is one way in which the two parts of logic are different. The algebra of statement connections has all the laws of the algebra of classes and it has one more law in addition.

6.4. Class and statement readings of Boole's algebra. At the start of his opening discussion in *The Mathematical Analysis of Logic* Boole says that those who are in touch with the present condition of 'the theory of Symbolical Algebra' are conscious of the fact that the validity of the processes of Mathematical Analysis is not dependent on the reading of the signs which are used but only on the laws by which the ways these signs may be put together are ruled. Any system of readings is as right and as possible as any other if, under it, all the laws of the theory become true statements about the material in connection with which the system of readings is offered. So the same process may, under one system of readings, be representative of the answer to a question on the properties of number, under another, to a question of geometry, and under a third, to that of a question of dynamics or optics.

Boole's algebra of logic is one theory; there are (at least) two systems of readings of it, one in connection with material about classes, the other in connection with material about statements. For example, in the algebra of logic under the class system of readings the sign 1 is All, this being the class whose elements are all the things that might be elements of the classes to be talked about, these classes getting their elements by selection of things from All; the sign 0 is Nothing, this being the class having nothing which is an element of

Boole's Algebra of Logic

All as an element; $x+y$ is the class whose elements are all the things of All which are elements of x or elements of y but not elements common to x and y; and $x \times y$ is the class whose elements are all the things of All which are elements of x and elements of y, so it is the class of things common to x and y, and if there are no such things, $x \times y = 0$. On the other hand, in the algebra of logic under the statement system of readings the sign 1 is The True, this being a statement by which all the possible events about which one may be talking are covered; the sign 0 is The False, this being a statement of an event which is not among the possible events; $x+y$ is 'x is true or y is true, but not the two of them together'; and $x \times y$ is 'x is true and y is true'.

For Boole a statement 'x' is separate from the statement ' "x" is true' in which 'x' is put forward as being true; the second statement is separate from the first if only because the first might be put forward as being false and not as being true. In Boole's algebra of statement connections the statement ' "x" is true' becomes the equation $x=1$, while the statement ' "x" is false' becomes the equation $x=0$. Because every x is true or false, a law of this algebra is: $x=0$ or $x=1$. This law is special to the algebra of statement connections (see end of 6.4).

6.5. Boole's 'Laws of Thought'. Boole's ideas about the algebra of logic were only outlined in *The Mathematical Analysis of Logic*. His complete statement of them was given in 1854 in his 424-page book *The Laws of Thought* (full name: *An Investigation of the Laws of Thought, on which are founded the mathematical theories of logic and probabilities*). (The other two of Boole's four books were not on logic but on higher branches of mathematics.) The first half of *The Laws of Thought* is on the algebra of logic, while the second half is on the uses of this algebra in the theory of probable reasoning and in the mathematics of chance. Most of the teachings in the first half are roughly the same as, though often given more fully than, those in his earlier book; together with them this time is a detailed account of the writer's opinions on their connections with and value for mathematics and philosophy. On a small number of quite important points, however, Boole's thought had undergone a change, sometimes for the better, sometimes not. Two such changes may be noted here. One is about the operation of division in his algebra; the other is about his views on the relations to time of a statement's being true or being false.

In *The Laws of Thought* Boole went very much further than he had done before in attempting to make use of the operations and processes of mathematics in building his theory of logic. It was his desire to let any idea of arithmetic or number algebra be used in his algebra of logic if it was a help to getting the right answers, even though some of the statements got from some of the processes of mathematics in the middle of the working out of a question in logic had no sense from the point of view of logic itself. Boole's belief was that his way was completely all right on condition that the last line of the working—the line giving the answer —did have sense for logic. An effect of this belief was that some of the rules of his algebra were guided by ideas in mathematics more than by the needs of a good theory of logic. No small number of steps in working out questions by his algebra are based on an operation of division where what is got by this has no sense for logic; and a specially strong protest is rightly to be made against his frequent use of signs, taken from what he gives the name of Arithmetic to (but which these days

40

Boole's Algebra of Logic

would be named Mathematical Analysis), like $\frac{0}{0}, \frac{0}{1}$ and $\frac{1}{0}$, whose senses and uses even in Arithmetic were then not very clear or free from error.

On the other point, about time, Boole says: let us make use of the letters X, Y and Z as marks of the simple statements on which we have the desire to put some value in relation to what is true and false, or among which we have the desire to put forward some relation in the form of a complex statement (such as 'X is true and Y is false' or 'if X is true then Y is true'). And let us make use of the small letters x, y and z in this way: let x be representative of an act of the mind by which one's looking at that stretch of time in which X is true is fixed; and let this be the sense to be given to the putting forward of x as the name of the time in which X is true. Let us further, Boole says, make use of the signs of connection $+$, $-$, $=$ and $\times$ in this way: let $x+y$ be the name of the group of those stretches of time in which X is true or Y is true, those times being completely separate from one another; let $x-y$ be the name of the rest of the time got when one takes away from the stretch of time in which X is true the stretch of time, which is a part of that first stretch, in which Y is true; let $x=y$ be the name of the statement that the time in which X is true is the very same as the time in which Y is true; and, lastly, let $x \times y$ be the name of the stretch of time in which the statements X and Y are true together. It is readily seen that such laws of number algebra as $x+y=y+x$, $x \times y=y \times x$ and $x \times (y+z)=(x \times y) +(x \times z)$ are true in this algebra of time: it is into an algebra of time that the algebra of statement connections has been turned by Boole, that is to say for the purpose of an account of the logic of statements he gave a time system of readings to his algebra of logic.

D   41

Boole's Algebra of Logic

6.6. Boole's algebra and number algebra. It is natural to put the questions: Are all the laws of Boole's general algebra of logic laws of the general algebra of numbers? and Are all the laws of the general algebra of numbers laws of Boole's general algebra of logic? The answer to the first question is, No. There is a law of Boole's algebra which is not a law of number algebra. This law is $x \times x = x$: in the algebra of classes it is true that the class of things common to x and to x is simply the class x itself and in the algebra of statement connections looked at from the point of view of times it is true that the stretch of time in which X is true and X is true is simply the time in which X is true, while in number algebra it is not true that $2 \times 2 = 2$ and, more generally, if x is any number greater than 1, it is not true that $x \times x = x$. And the answer to the second question is the same as to the first, No. The law of number algebra that if $x \times y = 0$, then $x = 0$ or $y = 0$ (and so the law, which is true if and only if that law is true, that when z is not 0, then $x = y$ if $z \times x = z \times y$) is not true in Boole's algebra of logic; for example, under the class system of readings, if All is the class of living persons, x is the class of mothers and y is the class of fathers, then $x \times y$ is equal to Nothing but x is not equal to Nothing and y is not equal to Nothing: though $x \times y = 0$, $x \neq 0$ and $y \neq 0$.

One of the chief points made by Boole in *The Laws of Thought* is that though his algebra of logic is not the same as the general algebra of numbers, it is the same from end to end as a more limited algebra, this being the algebra of the two numbers 0 and 1. The algebra of 0 and 1 is formed of the laws of algebra which are true when the variables used are representatives of only 0 and 1 (so, in fact, all the laws of high school algebra are laws of this algebra). Any one of the things of which a variable is a representative is said to be a 'value' of the

Boole's Algebra of Logic

variable; a variable in the algebra of 0 and 1 has only two possible values, 0 and 1. A necessary condition of Boole's algebra being the same in form as this algebra is that the somewhat strange law $x \times x = x$ which is true in the algebra of logic is true in the algebra of the numbers 0 and 1. The question is, then: Is the equation $x \times x = x$ true for the two possible values of x, that is, is it true that $0 \times 0 = 0$ and that $1 \times 1 = 1$? The answer is, Yes. But facing one necessary condition without a fall is not enough. To get full marks in a test the first question that is answered has to be answered rightly; however, being able to give the right answer to this part of the test paper is not all that is needed because, in addition, the rest of one's answers have to be right. Boole is in a like position; he has more troubles to overcome if his idea is going to get through our test without any loss. Let us now put another question: Is there a law of the algebra of 0 and 1 which is not a law of the algebra of logic? If there is such a law, Boole made an error in saying that his algebra was the same as that limited number algebra. There is such a law: when $z \neq 0$, $x = y$ if $z \times x = z \times y$. In the algebra of 0 and 1, when $z \neq 0$, $z = 1$, and if $1 \times x = 1 \times y$, then certainly $x = y$. In the algebra of logic, on the other hand, the implication 'if $z \neq 0$ and $z \times x = z \times y$, then necessarily $x = y$' does not have validity. The root reason why it does not have validity is that in the algebra of logic, when z is not 0, z is not necessarily 1. This is clear straight away from the class system of readings of Boole's algebra. Under this system of readings not every class is All or Nothing; there is a place for classes which have some elements of All without having all the elements of All as their elements. However, though it is not true that the general algebra of logic is the same as the algebra of 0 and 1, it is true that the algebra of statement connec-

43

tions is the same as that number algebra. The reader has seen that in the algebra of statement connections, for every statement x, $x=0$ or $x=1$: as in the algebra of 0 and 1, every variable has only two possible values, 0 and 1.

7

THE ALGEBRA OF LOGIC AFTER BOOLE: JEVONS, PEIRCE AND SCHROEDER

7.1. Later developments of Boole's algebra. Much of the work in Mathematical Logic in the 100 years and more that have gone by after *The Laws of Thought* has been given to taking out the errors from Boole's ideas, making some of the parts of his theory stronger, putting his algebra in the form of a system of deductions and, lastly, moving from it towards more general theories in Abstract Algebra, for example towards the theory of those 'part-ordered' classes every two elements of which have a greatest lower limit and a smallest higher limit, such part-ordered classes being what are named *lattices*. We will at this stage say something about one or two of the ways in which Boole's algebra was changed while he was still living or not long after his death; these changes were made by Jevons, Peirce and Schroeder.

7.2. The work of Jevons. William Stanley Jevons (1835–1882) was at the same time head of the school of

44

logic and philosophy and head of the school of economics at Owens College, Manchester (later the University of Manchester), from 1866 to 1876, going then, because of his heavy teaching work, to University College, London, as head of the school of economics. Like other men of science or letters under the rule of Victoria (for example Darwin and Clifford, Browning and Carlyle), Jevons was often ill, having frequent pains in the head and sleeping badly. So in 1880, when he was only 45 years old, he gave up his position at University College, living after that on his private income and the money from his books. But death came to him suddenly two years later while he was having a swim at Hastings. This event put an end to the writing of a full account of his teachings on economics. However, what had been produced by him earlier in this field had made his the chief name in current English economics. In addition, his was the chief name in the current English logic of deduction and logic of natural science; this was the effect of his six high quality books on logic.

The first of these books was *Pure Logic, or the Logic of Quality apart from Quantity* (1864); the greatest of them was *The Principles of Science* (1874) in which the writer's views on the logic of deduction and the logic of natural science were put, together, in their complete form.

One of Jevons's interests was in 'reasoning machines', of which he was the first to make one. Much of his time in the 1860's was given to the building of such an instrument by which the validity of deductions was able to be tested automatically.

In Jevons's opinion the theory of deduction is a theory of reasoning about qualities or quality-ideas and is not one of statements as such or of classes. The theory of deduction has to be one under whose rules the

The Algebra of Logic After Boole:

steps of any reasoning are turned into equations the signs on the two sides of which are representatives of clearly marked out qualities; certainly not turned into classes like the uncertain classes of which Boole made regular use in *The Laws of Thought*. 'All horses are animals' would be seen by Boole in *The Laws of Thought*, as it would have been seen by Hamilton, as saying 'all horses are some animals'; Boole's equation for this would be $h=v\times a$, h being the class of horses and a the class of animals, while v is a class unfixed in every way but this, that some of its elements are animals. But which of its elements are animals? Jevons's answer is that it is those which are horses. The right equation for 'all horses are animals' is $h=h\times a$, where 'h' and '$h\times a$' are names of qualities, one quality that is simple and one that is complex.

Jevons's theory was different from Boole's in another point. He had the idea that the heart of deduction is in the exchanging of signs, of things that are equal to one another, for one another. In his words, 'whatever is true of a thing is true of its like'. This idea was important later when one came to see that it is necessary to say what rules there are in a system by which deductions may be made. In the time of Boole and Jevons and for long after—till the early 1900's—no one, but for Frege, was conscious of the need to keep the laws in a system separate from its rules of deduction, which are for going from given laws to new ones.

7.3. Comparison of Boole's and Jevons's systems.
It was Jevons's belief that Boole went wrong in looking at logic as a branch of Mathematical Analysis and that his errors were the outcome of this. Every line in the working out of a question in logic has to have a sense from the point of view of logic. Boole lets a great num-

ber of lines be used though they have no such sense. Further, by Boole's account of the operation of addition $(+)$ of classes the operation may be done only on classes which have no common elements; however, by his rules he has to let '$x+x$' and '$x+(x+y)$' be parts of equations though x and x, and x and $x+y$, have common elements (if x and y are not Nothing). Again, while it is a general law of Boole's algebra that $x\times(y+z)=(x\times y)+(x\times z)$, it is possible for only one side of this equation to be well-formed, given Boole's account of $+$, because sometimes it is possible for the operation of addition to be done on $x\times y$ and $x\times z$, these two classes having no common elements, while it is not possible for the operation to be done on y and z, these two classes having common elements, as when 2 and 3 are the elements of x, 2, 4 and 5 are the elements of y and 3, 4 and 5 are the elements of z. Jevons put forward with much force the suggestion, which had been made before by De Morgan in connection with classes, that it would be better if the operation of addition in logic were made to be one for which there is no need of the things on which the operation is to be done to be completely different and to have nothing whatever common to them. In class algebra $x+y$ would then be the class of all the things that are elements of x or of y or of x and of y. One value of this suggestion is its causing number signs such as $2x(=x+x)$, which have no sense in logic, to be dropped out of the algebra, $x+x$ now being equal to x. With this change of addition Boole's operation of taking away $(-)$ is no longer the opposite of $+$ and is no longer like the $-$ of arithmetic or number algebra; $-x$ is made to be the class of all the things that are elements of All but are not elements of x, so now, for every class x, it will be true that $x+-x=1$, 1 being All, unlike in number mathematics where $x+-x=0$. Another value of Jevons's suggestion is that De Morgan's Laws, put forward by De Morgan for classes, these Laws being that $-(x+y)=-x\times-y$ and $-(x\times y)=-x+-y$, become true in the new theory, and this is a help in the answering of questions about reasoning.

Though Jevons's theory of deduction was better in some ways than Boole's, it was generally less good and more complex. The effect of this was that only two of his ideas were taken into the body of Mathematical Logic by later workers. One was the idea of changing the operation of addition. The other was his way of making an expansion of an equation to take more qualities or classes into account; for example, an expansion of $x=x$ is possible by which y is taken into account, as in $x=x=x\times(y+-y)=(x\times y)+(x\times-y)$. Such expansions were important for the later development of the idea of what was named the 'normal form' of a complex of signs, the normal form being a fixed way of writing any such complex so that comparisons between complexes and decisions of certain sorts about their properties may be readily made.

7.4. Peirce's points of agreement with Boole and Jevons.

Charles Sanders Peirce (1839–1914) was a teacher of philosophy, in logic, for only 5 years, at Johns Hopkins University in 1879–1884. He made his living for 28 years, from the time he got his degree from Harvard till 1887 when he was 48, from a branch of United States government work in science; after 1887 he had no regular income and was often so poor and in debt as not to have enough money for food and heating.

His ideas on Mathematical Logic were given in papers, or recorded in private notebooks, whose writing took place at widely different times between about 1866 and 1905, the most important years being those when he was at Johns Hopkins. Unlike Jevons, Peirce gave his approval to Boole's view of the logic of deduction as having close connections with mathematics, though he was in agreement with Jevons that any system of such a logic has to be limited in its use of mathematics by the need of letting all its laws and rules and all the steps which may be taken in it have good sense in logic.

There are four parts of logic that were much helped to go forward by Peirce: the logic of classes, of relations, of statement connections and of statements in which there are variables. These four will be touched on in turn.

7.5. Peirce's use of the 'a part of' relation. Peirce was the first worker in Boole's sort of logic to make use of the new operation of addition that had been put forward by De Morgan and Jevons. Down the years from 1867, the year of his first papers on Mathematical Logic, Peirce was responsible for a number of important developments and of new systems in the algebra of classes. Specially to be noted was his joining a new relation between classes to those which there had been in Boole's algebra. This new relation was 'a part of', which is currently marked by the sign $\subset$; the sense of $x \subset y$ is that every element of the class x is an element of the class y. In Boole's algebra thought is able to be given to this relation, which is one of the most important ones in logic and mathematics, only by an equation: $x \times (1-y) = 0$. But it is in fact a great help in logic and mathematics to have a simple sign for this relation. Further, there is this to be said for Peirce's use of the 'a part of' relation as a simple relation: this relation in the algebra of logic is like the relation of 'less than or equal to' in number algebra and, keeping this in mind, one may make new discoveries of laws of logic. If the algebra of classes is, in its form, like the algebra of the numbers 0 and 1, then, because there are true laws using the relation 'less than or equal to' in this number algebra, there are true laws using the relation 'a part of' in the class algebra. This was overlooked by Boole. Lastly, the sign ⊂ has the sense in the algebra of statement connections of material implication (2.8); when there was a separate sign for this statement connection, one was more readily able to give it attention and to make discovery of its properties. Peirce himself and others gave it much attention. Because of this, it became the generally used statement connection of the implication sort almost without a fight, even though material implication is a somewhat special and strange sort of implication. Its attractions are that it is the least strong sort of implication and that it is generally better than the others for the purposes of mathematics.

7.6. Peirce's theory of relations. Peirce gave a full account of his theory of relations in his paper 'The Logic of Relatives' printed in *Studies in Logic by Members of the Johns Hopkins University* (1883). His theory is based on Boole's algebra and on De Morgan's *Formal Logic* (1847) and papers of De Morgan. De Morgan put much weight on the idea of relations such as 'to the right of' which if they are true of the things *a* and *b* (in that order) and are true of the things *b* and *c* (in that order), then they are true every time of the things *a* and *c* (in that order): if *a* is to the right of *b* and *b* is to the right of *c*, then *a* is to the right of *c*. Further, De Morgan gave weight to the idea of an operation of joining two things together in thought so that they have a relation between them such as that between a head and a horse in what is marked by 'a head of a horse' or between a daughter and a teacher in what is marked by

50

'a daughter of a teacher'. (It may be pointed out that there are in fact a great number of different relations of this sort, though this is kept dark by the uncertain connection word 'of'.) It was De Morgan's belief that the 'of' relation was in some ways like the operation or relation marked by × in arithmetic or algebra; an example is the number law $a \times (b+c) = (a \times b) + (a \times c)$, which is, in addition, a law in the logic of relations, giving × the sense of 'of' and + the sense of 'or': the father of a teacher or a manager is the father of a teacher or the father of a manager. However, though relations were talked about by De Morgan and what he said had no small effect on Peirce, he did not get so far as building a system or a mathematics of relations; this is what Peirce made the attempt to do.

In 'The Logic of Relatives' an attempt was made by Peirce, using the work of Boole and De Morgan, to give a general theory like a mathematics of things having relations; these will be named here 'relation things'. The relation thing 'father' among persons may be said simply to be the class of all the groupings $(I:J)$ where I is a father and J is an offspring of I; so (James Mill: John Stuart Mill) is an element of the class that is the relation thing 'father' but (Roger Bacon: Francis Bacon) is not an element of that class. More generally, any relation thing r may be said to be the class whose elements are all the groupings $(I:J)$ of the things I and J such that, as one would commonly say, I is an r of J (or something near this). One value of this account by Peirce is its putting an end to the belief that a relation is something very strange and special. In this way Peirce did for relations what Boole had done for possible subjects and predicates: as qualities had been turned by Boole into classes of the things that have the qualities, relations were turned by Peirce into classes of the groups of things between which there are the relations. Another value is in its suggestion that ideas in mathematics which might be viewed at first as not a part of logic at all are able to be given *definitions* which make use only of what is a part of logic; mathematics and logic then are not quite as separate as it seemed, and it might even be possible for all the ideas of mathematics to be broken up into ones of logic, mathematics becoming an expansion of logic and no more than an expansion of it. (A definition is a statement or rule saying what sense some word or other sign has or is to have, or saying that some simple or complex sign is going to be used as equal to another sign or group of signs.)

7.7. Peirce's logic of statement connections. A logic of relations is needed for a logic of mathematics because there is a great number of relations playing an important part in mathematics, for example 'square root of', 'between', 'in the same plane as', 'greater than', 'equal to'. Relations are some of the material of mathematics. But the logic of statement connections is a theory of a part of the form, not of the material, of mathematics. Not that the business of these two sides of logic is only with mathematics—far from it. But much of their development, and that of other sides of logic, has in fact been caused by the desire of building a logic that is at least a good one for the needs and purposes of mathematics. There were chiefly three new ideas in the logic of statement connections for which Peirce was responsible. The first was the idea that it is possible to give definitions of all the different statement connections between one or two statements, as 'not', 'and', 'or' and the material implication 'if . . . then', by using the one connection 'the two are false' or by using the one connection 'not the two are true'. For example, 'the two are false' being marked by the writing of the statements s and t side by side so that 's and t are false' becomes 'st', then 's is false' becomes 'ss', 's is true or t is true' becomes '$(st)(st)$', and so on.

Secondly, Peirce gave an account of a 'decision process' for testing if a complex of signs in a system of statement connections is or is not representative of a true law of logic. A process is said to be a decision process if it is fully automatic—able to be done by machine—and gives a test of which complexes of signs in a system are laws of the system, a complex of signs being a law if and only if it has a certain property P. In Peirce's logic of statement connections the decision process is to give the values v and f (from the Latin *verum* = true and *falsum* = false) in all possible ways to the variables in the complex of signs and, by definitions, to get the value of the complex itself worked out from the values of its parts; if this value is necessarily v for all the possible values of the variables, then the complex is a law, while if this value may be f for some values of the variables, the complex is not a law; so the property P here is the property of a complex's having the value v for all possible values of the variables in it. Peirce makes this decision process clear in relation to the complex C that is of the form $(s \to t) \to [(t \to u) \to (s \to u)]$, where $\to$ is a sign of material implication. By definition, $A \to B$ has the value v whatever values v and f the parts A and B of a complex might have but for A having the value v and B having the value f, when $A \to B$ has the value f: a material implication is false if and only if the condition is true and the outcome is false. Now Peirce's argument is this, that if C ever has the value f, then, from the definition, $s \to t$ has the value v and $(t \to u) \to (s \to u)$ has the value f; but if this last has the value f, then, from the definition, $t\rightarrow u$ has the value v and $s\rightarrow u$ has the value f; and so, again from the definition, s has the value v and u has the value f. Because u has the value f and $t\rightarrow u$ has the value v, necessarily, from the given definition of material implication, t has the value f. But then, because s has the value v and t has the value f, $s\rightarrow t$ necessarily has the value f. This, however, is not in harmony with $s\rightarrow t$ having the value v when C has the value f (10 lines back). So it is not possible for C to have the value f at all: if it is given that value, one gets values that are not in harmony with the definition of material implication. In ways like this all complexes of signs in a system of statement connections may be tested to see if they may be given the value f without effecting values that are not in harmony with some definition, in relation to v and f, of a statement connection.

Thirdly, Peirce had the idea of building up a theory of the logic of statement connections as a system based on axioms. Because the deduction of theorems about statement connections from axioms is very hard, it was Peirce's suggestion that the decision process we were outlining a minute back be used for supporting the right of those complexes which are unable to have and to keep the value f to be theorems in the system and so representatives of laws of the logic of statement connections. The validity of any argument in mathematics which is dependent on statement connections is conditioned by these laws: if the argument's validity is dependent only on statement connections, then the argument has validity if and only if it is in agreement with these laws; if the argument's validity is dependent on statement connections and on other sorts of reasoning, then agreement with those laws is necessary but not enough to make certain of validity. Peirce's system has four axioms in which material implication is the only statement connection and one axiom whose purpose seems to be to give a definition of the idea of 'not' or 'false'.

7.8. Peirce's quantifier logic. The statement '4 is the square of 2' is true or false, as is 'the death of Newton took place in 1900'. We will say that such statements are two-valued ones, the values being 'true' and 'false'. On the other hand, 'x is the square of y', where the range of the variables is the class of the numbers 1, 2, 3, . . . , and 'the death of x took place in y', where the range of x is the class of persons and the range of y is the class of years in the current system, are statements that are not two-valued; they are not true and they are not false. We will say that a statement which is not two-valued and in which there are variables is a statement with free variables. $s(x)$, $t(x, y)$, $u(x, y, z)$ and so on are used for noting such statements; for example $s(x)$ is the name of any statement having one variable in it. (This is a little over-simple; it is unnecessary here to go into delicate details, because we are having only a bird's-eye view of logic.)

Looking at the statements 'some x is the father of some y', 'some x is the father of every y', 'every x is the father of some y' and 'every x is the father of every y', the range of the variables being the class of persons, one sees that all four of them are two-valued; the first is true and the others are false The first statement says that someone is somebody's father, the second that somebody is the father of everyone, the third that everyone is somebody's father and the fourth that everyone is the father of everyone. Putting these four statements in a somewhat different form the first becomes 'there is an x and there is a y such that: x is the

The Algebra of Logic After Boole:

father of y', the second becomes 'there is an x such that for every y: x is the father of y', the third 'for every x there is a y such that: x is the father of y' and the fourth 'for every x and for every y: x is the father of y'. Though the part coming after the : in these four statements is a statement with free variables, the complete statements themselves are two-valued. 'There is a' and 'for every' are named *quantifiers*, the first being the 'existence' and the second the 'general' quantifier, which were noted by Peirce by the Greek letters Σ and Π.

Peirce gave much attention to the logic of variables, this being that field of logic whose business is with implications and arguments some or all of whose conditions or outcomes are statements with free variables or that have quantifiers in them. A great number of theorems and arguments in mathematics are dependent on the use of quantifiers; so this part of logic is very important. With Frege but having no knowledge of Frege's work, Peirce was the first to be interested and to make deep discoveries in this field. Using — as the sign of 'not' or 'it is false that', some simple examples of laws given by Peirce are:

(1) $-(\Sigma x)s(x)=(\Pi x)-s(x)$,
(2) $-(\Pi x)s(x)=(\Sigma x)-s(x)$,
(3) $(\Sigma x)(s(x)+t(x))=((\Sigma x)s(x))+((\Sigma x)t(x))$,
(4) $(\Pi x)(s(x)+t(x))=((\Pi x)s(x))+((\Pi x)t(x))$,
(5) $(\Sigma x)(\Pi y)u(x, y)\rightarrow(\Pi y)(\Sigma x)u(x, y)$,

and laws which are the same as (3) and (4) but in which $\times$ takes the place of $+$.

7.9. Schroeder's algebra of logic.

Ernst Schroeder's (1841–1902) two chief books on Mathematical Logic were *Der Operationskreis der Logikkalkuls* ('The Range

56

of Operation of the Mathematics of Logic') and *Vorlesungen über die Algebra der Logik* ('Teachings on the Algebra of Logic'). The first of these two books (1877) is very short, being only 37 pages long. There is in it a simple system of the algebra of logic, this system being based on Boole's though different from it in some important points where the discovery had been made by later thought and experience that changes were wise or necessary. The tendency of these changes was in the same direction as that of those changes that had been put forward by Jevons and Peirce: the algebra is to be logic in the dress of mathematics, not a bit of mathematics which almost by accident is open, completely or in part, to being viewed as a theory of logic.

The writing of the *Vorlesungen* was in four parts that came separately on the market between the years 1890 and 1905. This book gave a full account of the current Mathematical Logic in so far as this had its roots in algebra; in it were discussions and systems going into the smallest details of the logic of classes, of statement connections and of relations.

Much that Schroeder did was new, though often resting on older material. Among the number of his theorems that were of special value was that to the effect that all the laws of the algebra of logic may be grouped in twos, it being possible from a knowledge of one of the two to get straight away, by a simple rule of exchange, a knowledge of the other one. The rule is that, in the given law, $+$ and $\times$, and 1 and 0 (the signs of All and Nothing), be put in one another's places; for example, from the law $x \times 1 = x$ one may get the law $x + 0 = x$, from the law $x + -x = 1$ one may get the law $x \times -x = 0$ and from the law $x \times (y+z) = (x \times y) + (x \times z)$ one may get the law $x + (y \times z) = (x+y) \times (x+z)$. This last law was first given by Peirce; it is not a law in Boole's

B   57

algebra because of the different use Boole makes of $+$.

Again, the laws $x+x=x$, $x+y=y+x$, $(x+y)+z=x+(y+z)$ and $x=x+(x\times y)$ and the four laws that may be got from these by the rule of exchange were seen by Schroeder to be of a somewhat separate sort from the other laws. In fact, these 8 laws are very important for Lattice Theory (see end of 7.1) because a class L is a lattice if and only if these laws are true in it.

7.10. Whitehead's and Huntington's work on the algebra of logic. The present history of the algebra of logic will be ended now by our saying a word or two about the stage in the development of this algebra that came in the ten years after its first 50 years, 1847–1897. This stage is marked by two things. Firstly, by Whitehead's (1861–1947) attempt in his *Universal Algebra* (1898) to give a completely general theory of algebra and to say what bodies of axioms and definitions will be good starting points for the different systems covering this part of mathematics; this book was the first after *The Laws of Thought* to take seriously the algebra of logic as a field of everyday mathematics and it was the first to make clear the connections between this branch and other branches of Abstract Algebra. Secondly, this stage is marked by the first conscious work in the metamathematics of the algebra of logic (see 5.2). Using tricks and processes of a sort designed by Peano and by others of the Italian school, Huntington (1874–1952) gave a *demonstration*—a good argument in support of what is said to be true—of the consistency of a number of different axiom systems of the algebra of logic, the substance of these systems being copies of ones in the books of Schroeder and Whitehead, and he gave a demonstration that every axiom of his systems is unable to be got as a theorem from the rest of the axioms of the same system, so no axiom of a system is dependent on the others. This bit of second-order theory of the algebra of logic—the logic of logic—was done at the time (1904), between 1890 and 1905, when second-order theories of mathematics—metamathematics—were starting to be seen as interesting and as a very important part of the theory of the bases of mathematics.

8

FREGE'S LOGIC

8.1. The mathematics of logic and the logic of mathematics. Mathematical Logic has two sides. One, in its roots in the past (we say this because all sciences have a tendency to become self-conscious and, if necessary for their best development, to let the material that was their first business get dropped from view), is the science of the deduction form of reasoning, this science using ideas and operations like those in mathematics and copying mathematics in being system-making. This side, then, is the mathematics of logic. It was in this that Boole and others interested in the algebra of logic were working. This side has two levels: the logic of deduction itself in the form of systems covering statement connections, classes, relations and statements with variables; and the second-order theories of those systems, whose birth came nearly 50 years after that of those first-order systems. The other side of Mathematical Logic is the logic of mathematics. Here there are two chief sorts of questions to which one has given attention, certain points of these being in the same line of direction. One sort of question is that to be answered by the metamathematics of any given branch of mathematics—of arithmetic, Euclid's geometry, group theory or whatever it may be. The second sort of question is about the bases of mathematics; for example, is it possible to have a system with the property of consistency in which the ideas used are enough and have the right properties so that, firstly, definitions of all the ideas of the common mathematics may be given from these and only these ideas, and, secondly, all the theorems in the common mathematics may be got as theorems from that system stretched to have new definitions resting on the ideas in the old ones? From the time of Schroeder's death (1902) till now, most work in Mathematical Logic has been in connection with the logic of mathematics, and in 8.2–12.6 we will say something about this work. The reading of this coming discussion will have to be done very slowly if the material is going to be taken in. Because we have little space and because it is possible to get a clear view of what has been produced by the newer work in Mathematical Logic only if one goes through the long stretch of expert details, we will be limiting ourselves to outlining some of the chief directions which the later developments in Mathematical Logic have had.

8.2. Reasons why Frege's writings were undervalued. Gottlob Frege (1848–1925) was Professor of Mathematics at the University of Jena and the writer of books and papers on logic and the logic of arithmetic, and on the philosophy of these fields. For a number of reasons Frege's writings were undervalued till Russell's *Principles of Mathematics* (1903) gave an account of them. One reason was their use of signs for statement connections the writing of which was not in the normal direction of left to right but up and down; this makes the reading of sign complexes much harder. Another reason was that, generally, those persons whose business was philosophy were controlled by the impulse not to go near, still less to go through, a mass of sign complexes when the purpose of this mass was to be the support of a theory in philosophy—here, in the philosophy of mathematics. Their belief was (and is) that an argument for or against any such theory has to be made in everyday language because the value of the theory is dependent on its starting points, and the opinions which are at the back of these, being true, and that it is not possible for these to be supported by anything inside a mathematics-like structure but only by arguments outside the structure. Another reason was that Frege's theory was so surprising and so far from probable that there seemed little chance of the trouble of working through the details of the deductions offered for it being rewarding. One last reason we would put forward is that what Frege was doing was off the straight and narrow road: it was not quite normal mathematics, it was not quite normal philosophy and it was not quite normal logic. It was not normal mathematics: it was the bases of mathematics. It was not normal philosophy: it was the philosophy of mathematics done by someone having a first-hand knowledge of mathematics. And it was not normal logic: it was not the algebra of logic. (The 'normal' is in relation to the time 1875–1900.)

8.3. Frege's view of the relation between logic and arithmetic. What was Frege's theory? It was that number is an idea (not dependent on the mind for its existence) which may be broken up into ideas of logic and which is open to a complete definition by using no

61

more than these ideas; and so, further, that arithmetic itself is a part of logic. In agreement with Frege, Russell said that arithmetic 'is only a development, without new axioms, of a certain branch of general logic'. The journey to the full, tightly reasoned demonstration of this theory was started by Frege in his *Begriffsschrift; eine der arithmetischen nachgebildete Formelsprache des reinen Denkens* ('Idea-Writing; a sign language, copying arithmetic, of thought as such') of 1879 and came to its end with the two parts of his *Die Grundgesetze der Arithmetik, begriffsschriftlich abgeleitet* ("The Root Laws of Arithmetic, got by deductions in idea-writing') of 1893 and 1903. In the *Begriffsschrift* Frege gives an axiom system of the logic of statement connections and, using the general quantifier (by which, together with 'not', he gives a definition of the existence quantifier), he goes deeply into the logic of statements with variables. In addition, there is an account of some parts of arithmetic worked out from definitions based on what are taken to be ideas of logic, for example the ideas of class, class element and relation. In the *Grundgesetze* there is a complete account of arithmetic, in so far as this is the theory of the laws of the operations on numbers of the form p/q where the range of q is $\pm 1, \pm 2, \pm 3, \ldots$ and the range of p is the same together with 0. The heart of Frege's arithmetic is his definition of a *cardinal* number. The simplest cardinal numbers are 0, 1, 2, 3, ... For Frege, a cardinal number is a property of a class. To the question, 'What is the number of men in the Russian army now?' the right answer will be a cardinal number and this number will be a property of the class 'man in the Russian army now'. By a 'class' here Frege, unlike Russell, has in mind the class-idea (that is, the class as an idea), not the list of things coming under the class-idea. So a cardinal number is a property of a class-idea, for example of the class-idea 'man in the Russian army now'. Frege's definition of the cardinal number of a class x is that it is the class of all the classes y such that x and y have a one-one relation between them; by definition, classes x and y have a one-one relation between them if and only if (i) each element of x may be grouped with some element of y, (ii) each element of y may be grouped with some element of x, (iii) if any a_1 and a_2 of x are grouped with a certain b of y, then a_1 and a_2 are the very same element of x, and (iv) if a certain a of x is grouped with b_1 and with b_2 of y, then b_1 and b_2 are the very same element of y. This may seem harder than it is; it is put like this for the purpose of stopping anyone from being able to say that the definition goes round in a circle, because it makes use of the cardinal number one (in 'one-one relation'). What the definition says is simply this, that x and y have a one-one relation between them if and only if it is possible for there to be a body R of groupings $(a : b)$ such that every a of x is the first name in some grouping $(a : b)$ of R while no a of x is the first name in more than one such grouping and every b of y is the second name in some grouping $(a : b)$ of R while no b of y is the second name in more than one such grouping. From his definition of cardinal number and by the instruments of logic Frege gets the theorems of everyday arithmetic at the end of very long chains of deduction in the form of sign complexes; so even if one has doubts of his theory being true, at any rate his industry and expert invention are to be respected.

8.4. Frege on functions. One of the most important ideas at the base of mathematics is that of a *function*. The definitions given of this idea in Frege's time did not, when looked at with care, make very good sense though they had seemed to make good sense and, further, they were not in harmony with the uses made of the idea. As frequently in philosophy, Frege had a sharp tongue when attacking, specially when attacking an unsafe position.

What is a function? In Frege's time the common answer to this question was to say, as the German writer on mathematics, Ernst Czuber, did: 'If every value of the number variable x that is an element of the range of x is joined by some fixed relation to a certain number y, then, generally, y, like x, comes under the definition of a variable and y is said to be a function of the number variable x. The relation between x and y is marked by an equation of the form $y=f(x)$.' (Number variables are not necessarily limited to the range 0, ±1, ±2, ±3, . . ., or even to numbers of the sort p/q where p and q are from that range ($q\neq0$); they may, for example, be variables of numbers like $\sqrt{2}$ or $\sqrt[3]{-25}$.) This answer was soon undermined by Frege's blows, one after another, from the gun of his quick-firing brain. 'One may straight away take note of the fact', says Frege, 'that y is named a certain number, while, on the other hand, being a variable, it would have to be an uncertain number. y is not a certain and it is not an uncertain number; but the sign 'y' is wrongly placed in connection with a class of numbers [the range of the variable y] and then it is talked about as if there were only one number in this class. It would be simpler and clearer to say: "With every number of an x-range there is some fixed relation by which it is joined to another [not necessarily different] number. The class of all these last numbers will be named the y-range." Here we certainly have a y-range, but we have no y of which one might say that it was a function of the number variable x.

Frege's Logic

'Now the limiting of the range does not seem to have anything to do with the question of what the function itself is. Why not take as the range the class of all the numbers of arithmetic or the class of all the complex numbers (of which the first class is a part)? The heart of the business is in fact in quite a different place which is kept out of view in the words "joined by some fixed relation". What is the test of the number 5 being joined by some fixed relation to the number 4? The question is unable to be answered if it is not somehow made more complete. With Herr Czuber's account it seems as if, for any two numbers, the decision was made at the start that the first is, or is not, joined by the relation to the second. Happily Herr Czuber goes on to say: "My definition makes no statement about the law controlling the joining by the relation; this law is marked in its most general form by the letter f but there is no end to the number of different ways in which it may be fixed." Joining by a relation, then, takes place in agreement with some law and it is possible, by the use of thought, for different laws of this sort to be produced. But now the group of words "y is a function of x" has no sense; there is need of its being made complete by the addition of the law. This is an error in the definition. And, without doubt, the law, which is overlooked by this definition, is truly the chief thing. We see now that what is variable [changing] has gone completely out of view and in its place the general has come into view, because that is what is pointed at by the word "law".'

8.5. Functions as rules. What Frege says about functions in the paper from which we have taken these lines has become normal in the present-day account of the idea of a function. A function f is a rule about classes X and Y, X being named the class of definition

65

of f and Υ the range of f, the rule producing a class of ordered groupings $(x : y)$ where x (which is named an 'argument' of f) is an element of X and y (which is named the 'value' of f for x) is that selection from among the elements of Υ which is made in agreement with the rule; the class of all the selections y from Υ is named the class of values of f. For example, if the class of definition of f is the class X of all the numbers 1, 2, 3, ..., the range of f is the same class and f is the rule 'To every element x of X let $2x$ be joined', then the class of ordered groupings produced by this rule has $(1 : 2)$, $(2 : 4), (3 : 6), \ldots$ as its elements, so the class of values of f has 2, 4, 6, ... as its elements; again, if the class of definition of f_1 is the class X of all the numbers -1, -2, -3, ..., the range of f_1 is the class of all the numbers 1, 2, 3, ... and f_1 is the rule 'To every element x of X let x^2 be joined', then the class of ordered groupings produced by this rule has $(-1 : 1), (-2 : 4), (-3 : 9), \ldots$ as its elements and the class of values of f_1 has 1, 4, 9, ... as its elements.

9

CANTOR'S ARITHMETIC OF CLASSES

9.1. Finite and enumerable classes. A short way of writing that a one-one relation is possible between any two classes x and y is '$x \approx y$'. If $x \approx y$, then by a definition of Cantor's (1845–1918), x and y are said to have the same cardinal number or the same power, the sign

Cantor's Arithmetic of Classes

for this fact being the equation '$\bar{\bar{x}}=\bar{\bar{y}}$'. (This definition is clearly like Frege's definition of 'cardinal number'; but Cantor does not give a definition of 'cardinal number', only of 'having the same cardinal number'. Cantor and Frege were working at the same time, but quite separately; one was not copying the other at all.) If N_m is the class of the 'natural' numbers from 1 through m, so the elements of it are 1, 2, 3, ..., $m-1$, m, then a class x is *finite* if and only if it has no elements or, for some natural number m, $x \approx N_m$. If N is the class of all the natural numbers, its elements being 1, 2, 3, ... where '...' is used for marking that the list goes on for ever, then a class x is said to be *enumerable* if and only if $x \approx N$. It is strange how great a number of quite different classes are enumerable. For example, Cantor gave demonstrations that the class Z of all the numbers $0, \pm 1, \pm 2, \pm 3, \ldots$ is enumerable; the class of all the numbers p/q greater than 0 and less than 1, where p and q are elements of N, is enumerable; but what to the mind's eye would be seen as a very much greater class, the class Q of all the numbers p/q, where p and q are elements of Z ($q \neq 0$), is again enumerable; and, further, the class of all numbers which are roots of an equation of the form $a_n w^n + a_{n-1} w^{n-1} + \ldots + a_1 w + a_0 = 0$, where each a_i is an element of Q, is enumerable.

In view of these surprising facts it is natural to put the question: Are there any classes that are unenumerable, that is which are not finite and which are unable to be placed in a one-one relation with N? The answer is, Yes. The class I of all the numbers of the form $0 \cdot a_1 a_2 a_3 \ldots a_n \ldots$, where the a's have values in the range 0, 1, 2, ..., 9, is an unenumerable class; it has, for example, $0 \cdot 127845 \ldots$, $0 \cdot 23640000 \ldots$ and $0 \cdot 765976 \ldots$ as elements. There are other unenumerable classes.

Cantor's Arithmetic of Classes

9.2. Cantor's cardinal arithmetic. Though Cantor did not give a definition of 'cardinal number' he made use of cardinal numbers as things having some sort of existence and having properties and relations among themselves, in much the same way as one makes use of the numbers 1, 2, 3, . . . in everyday arithmetic without having definitions of them. By definition, the cardinal number of a class x is greater than the cardinal number of a class y —$\bar{\bar{x}} > \bar{\bar{y}}$—if and only if there is some part x_1 of x such that x_1 is not all of x, $x_1 \approx y$ and there is no part y_1 of y such that $y_1 \approx x$; for example, $\bar{\bar{N_3}} > \bar{\bar{N_2}}$ and $\bar{\bar{N}} > \bar{\bar{N_{100}}}$. Cantor gave a demonstration of a general theorem which says that, for any class y having elements, the class Uy whose elements are all the classes x such that $x \subset y$ (7.5) has a greater cardinal number than y has: $\bar{\bar{Uy}} > \bar{\bar{y}}$. An outcome of this theorem is that there are ever greater and greater cardinal numbers, if only because $\bar{\bar{y}} < \bar{\bar{Uy}} < \bar{\bar{UUy}} < \bar{\bar{UUUy}} < \bar{\bar{UUUUy}} < \ldots$

An arithmetic of cardinal numbers was worked out by Cantor, definitions of the operations of addition and so on for cardinal numbers being given and the discovery and demonstration made of laws of the arithmetic. To take the simplest operation, that of addition: by definition, if x and y are any two classes such that no element of the one is an element of the other, then $\bar{\bar{x+y}} = \bar{\bar{x}} + \bar{\bar{y}}$, where the $+$ sign in the left-hand side of the equation is of the addition of classes ($x+y$ being the class whose elements are all the things that are elements of x or of y). The common laws of addition are true for cardinal numbers; if k_1, k_2 and k_3 are cardinal numbers, $k_1+k_2=k_2+k_1$ and $(k_1+k_2)+k_3=k_1+(k_2+k_3)$. On the other hand, there are, further, some uncommon laws;

for example, if m is the cardinal number of any finite class and k is the cardinal number of N, then the addition of m k's is equal to only k ($mk=k$) and $k^m=k$.

9.3. Cantor's ordinal arithmetic. The arithmetic of still another sort of number, that named *ordinal*, was worked out for the first time by Cantor. Ordinal numbers are based on well-ordered classes, that is on classes which are made to have a certain order among their elements so that there is a first element, a second element, and so on. For finite classes ordinal and cardinal numbers are the same; however, for classes that are not finite the numbers are not the same because it is possible to make all such classes have widely different order relations. And, as for cardinals, there is no limit to the number of higher and higher ordinals.

The theory of cardinal and ordinal numbers—the arithmetic of classes—has been one of the most fertile inventions of the newer mathematics and is one of the most important branches of Mathematical Logic. The details of this theory are highly complex, and we are unable to go into them here.

9.4. Burali-Forti's and Russell's discoveries. But the warm bright days when the sunlight was playing on the flowers in Cantor's garden and all seemed well were sadly soon over. Black clouds came up in the sky and the earth was made dark. Men of mathematics were put in doubt, fearing to see the downfall and destruction of one of the beautiful theories of mathematics.

The cause of the trouble was the discovery that Cantor's arithmetic of cardinal and ordinal numbers did not have the needed property of consistency (5.1). In 1897 Burali-Forti (1861–1931) gave a demonstration

Cantor's Arithmetic of Classes

that the ordinal number of the well-ordered class W of all the ordinal numbers is not an element of that class; so, on the one hand, all ordinal numbers are, by definition, elements of W while, on the other hand, not all ordinal numbers are elements of W because the ordinal number of W itself is not an element of W. Again, in 1901 Russell (1872–) saw the loss of consistency of cardinal arithmetic in connection with the most general class A. By definition, A is the class whose elements are all the things which are not classes, together with all the classes of those things, together with all the classes of classes of those things, and so on. A, then, is a class such that all its parts are elements of it. Because of this fact, it is not possible for there to be more parts of A than there are elements of A, that is $\overline{\overline{UA}} \leqslant \overline{\overline{A}}$ (the cardinal number of the class of all the parts of A is less than or equal to the cardinal number of A). However, this is not in agreement with Cantor's general theorem (9.2) which makes it necessary here that the cardinal number of UA is not less than or equal to, but greater than, the cardinal number of A. From the discovery of this example Russell got the idea of another one which is the most noted of all such examples. He sent it to Frege at the very time that the second part of Frege's *Grundgesetze der Arithmetik* was coming near the end of its printing before being offered to the public; Frege was made conscious of the fact that his system of classes, like Cantor's, did not have the needed property of consistency. Russell's new idea was this: without there being any suggestion that any class is able to be an element of itself, let E, by definition, be the class of all classes that are not elements of themselves; for example, the class 'man' is an element of E because 'man' is not an element of itself, this class having only men, and no classes, as elements. Is E an

70

Cantor's Arithmetic of Classes

element of E? If (I) E is an element of E, then (II) E is not an element of E because the only elements of E are those classes which are not elements of themselves; on the other hand, if (II) E is not an element of E, then (I) E is an element of E because of the definition of E. By a common law of logic one or the other of (I) and (II) is true. But whichever one is true makes the other one true in addition. So the theory of classes has two opposite theorems in it: E is an element of E; E is not an element of E.

The effect of these discoveries on the development of Mathematical Logic has been very great. The fear that the current systems of mathematics might not have consistency has been chiefly responsible for the change in the direction of Mathematical Logic towards meta-mathematics, for the purpose of becoming free from the disease of doubting if mathematics is resting on a solid base. A special reason for being troubled is that the theory of classes is used in all other parts of mathematics; so if it is wrong in some way, they are possibly in error. Further, quite separately from the theory of classes, might not discoveries of opposite theorems in algebra, geometry or Mathematical Analysis suddenly come into view, as the discoveries of Burali-Forti and Russell had done? It has been seen that common sense is not good enough as a lighthouse for keeping one safe from being broken against the overhanging slopes of sharp logic. To become certain with good reason that the systems of mathematics are all right it is necessary for the details of their structures to be looked at with care and for demonstrations to be given that, with those structures, consistency is present.

This last point and the fears and troubled mind we have been talking about in these lines have been and are common among workers in what is named 'the bases of mathematics', that is axiom systems of logic-classes-and-arithmetic. However, some persons, with whom the present writer is in agreement, have a different opinion. They would say that the well-being of mathematics is not dependent on its 'bases'. The value of mathematics is in the fruits of its branches more than in its 'roots'; in the great number of surprising and interesting theorems of algebra, analysis, geometry, topology, theory of numbers and theory of chances more than in attempts to get a bit of arithmetic or topology as simply a development of logic itself. They would say that the name 'bases of mathematics' is a bad one in so far as it sends a wrong picture into one's mind of the relations between logic and higher mathematics. Higher mathematics is not resting on logic or formed from logic. They would say that the troubles in the theory of classes came from most special examples of classes, and such classes are not used in higher mathematics. They would say further that though to be certain of consistency is to be desired if such certain knowledge is possible to us, a knowledge of the consistency of the theories of mathematics that is probable is generally enough and the only sort of knowledge of consistency that one does in fact generally have. And they would say that such probable knowledge is well supported if the theories of mathematics have been worked out much and opposite theorems in them have not come to light. There is no suggestion in all this that Mathematical Logic is not an important part of mathematics; the view put forward is that there is much more to mathematics than Mathematical Logic is and might ever become. And there is no suggestion that the question of consistency and like questions, and the discovery of ways of answering them, are not important; to no small degree Mathematical Logic now is as interesting and important as it is because of its interest in such questions and answers.

10

PEANO'S LOGIC

10.1. Implication and mathematics. In its birth Mathematical Logic was the theory of classes. The first person to have the opinion that the theory of statements (with or without variables) is more important was Hugh McColl (1837–1909) who in a group of papers on 'The calculus of equivalent statements' (1878 and later) put forward the belief that the only business of logic is with the theory of statements and that the chief statement connection is some sort of implication. The thought that the theory of statements and not that of classes is the root of Mathematical Logic, and the thought that implication of one sort or another is the chief relation to be given attention in logic, soon became the normal beliefs of those at the head of this field. For example, Frege and Peirce were interested in the logic of statement connections as a branch of logic separate from the algebra of class logic and implication was specially important in their systems. Before Peano (1858–1932), however, no one made use of the logic of statements for making clear the arguments of everyday mathematics, and so viewing logic as an instrument for getting clear and tight reasoning in such mathematics. (Peano was the first to give the new logic the name of 'Mathematical Logic', because of his view of it as an instrument for mathematics: for him, Mathematical

Peano's Logic

Logic is the logic of mathematics.) And it had not been pointed out before Peano that implication is the chief relation in mathematics, all or almost all the statements that are true in any system of mathematics being implications. (It was because of this teaching of Peano's about the theorems of mathematics being implications that Russell, in the opening of his *Principles of Mathematics*, gave as his definition of mathematics that this was the class of all statements of the form 'if s_1, then s_2', s_1 and s_2 being limited in certain ways.) And the idea that it was possible, by the use of logic, for all the statements in mathematics, and not only in arithmetic, to be put in the form of a language of made-up signs and for the demonstrations of all its theorems to be done by changes and exchanges of these signs, starting from axioms and definitions as sign complexes, had not been acted on in detail before Peano. The other experts in Mathematical Logic at that time were interested in logic for itself or were interested, as Frege was, in turning a bit of mathematics into logic; unlike Peano they were not interested in the value of logic as an instrument for everyday mathematics as everyday mathematics (and not as logic or physics or philosophy).

10.2. Peano's purpose in his logic. Peano's purpose in Mathematical Logic was to make the demonstrations in mathematics tight and free from loose reasoning. His view was that the value of what is offered as a demonstration and its being a good or a bad argument are not dependent on taste or inner feelings but on the argument's having the property of validity which is publicly testable. Peano said that because the old logic is not of much use in mathematics, the arguments in this not being syllogistic, it has been judged by some, Descartes among them, that what is clear to the mind as being

Peano's Logic

true is the only test of an argument's being right. He gives the words of Duhamel: 'La déduction se fait par le sentiment de l'évidence, qui n'a besoin d'aucune règle, et ne peut être supplée par aucune—A deduction is made by a feeling of what is clear to the mind as being true, which has no need of any rule and is unable to be given by any.' Peano's views about demonstration were a reaction to views of this sort.

10.3. Some of Peano's discoveries and inventions.
To put the demonstrations of mathematics in a tightly reasoned form Peano undertook the discovery of all the ideas and laws of logic that are used in mathematics and undertook the invention of a body of signs for clearly noting these ideas and for the clear statement of these laws. Among his discoveries and inventions were these: (1) the definition of a class by a statement of the form 'the class of x's such that $p(x)$', this form being marked by Peano by '$\overline{x\epsilon}\ p_x$'; (2) the idea that statements with free variables are different in important ways from two-valued statements; (3) the use of full stops, in the place of signs such as (,) and [,], for grouping complexes of signs; for example, the sense of $a \vee b . = \therefore \sim\ :\sim a . \sim b$ is ' "a or b" is equal to "not-((not-a) and (not-b))" '; (4) the use, for the operations and relations of logic, of signs that are unlike those of mathematics where there might be danger of a wrong reading; (5) the first pointing out of the fact (which Schroeder and others had not been conscious of) that the relation of being an element of a class is very different from the relation of being a part of a class, Peano noting the first relation by ϵ and the second by the 'horseshoe' $\supset$ (which became the normal sign for material implication after its use for that purpose by Whitehead and Russell in their *Principia Mathematica*); (6) the idea of 'the so-and-so', noted by

75

ιx, this idea being needed in connection with properties of which one has the desire to say that there is one and only one thing having them; (7) the noting of the general quantifier (7.8) by writing the variables under and to the right of the statement connection, so that for example the sense of $x \in a. \supset_x . x \in b$ is 'for every x, if x is an element of a, then x is an element of b'; (8) the noting of the existence quantifier 'there is a' by E printed back to front, that is by Ǝ. Most of these and no small number of other signs of Peano's became normal ones in Mathematical Logic after their use by Whitehead and Russell.

10.4. Peano's school of metamathematics. More than logic itself is needed for getting tightly reasoned demonstrations in mathematics. It is necessary for there to be complete lists of axioms and definitions: there has to be a full, clear statement of one's starting points in any system. Peano did much important work in this field of axioms and definitions. But Peano's interest in this work was guided by questions of logic: metamathematics. He was interested in questions like this: What is the smallest number of ideas and axioms, and which ones are they, that are necessary for working out an axiom system covering such-and-such material? And how is it possible to give reasoned answers to such questions?

Round Peano in Italy was formed a company of those experts who were interested in the axiom bases of mathematics and in the use of a designed language of signs for the theorems and arguments of mathematics. This group headed by Peano had control of a paper *Rivista di Matematica* (1891–) and were joined in the writing of a book *Le Formulaire de Mathématiques* (first printing 1894–95, last printing 1908 when the language used in it other than the signs for the material of mathematics was Peano's international language Latino Sine Flexione) for the detailed development of their theories.

11

WHITEHEAD AND RUSSELL'S *PRINCIPIA MATHEMATICA*

11.1. Short account of the three parts of 'Principia Mathematica'. The high-water mark of the first stages in the growth of Mathematical Logic, the stages from 1847 to 1910, was Whitehead and Russell's *Principia Mathematica* (first part, 1910; second part, 1912; third part, 1913). This is a great book, great in quality and great in size—its three parts have more than 2,000 pages. In it are united into one general system the Boole-Schroeder algebra of logic and the theories, differently purposed from that and from one another, of Frege, Cantor and Peano.

After an opening discussion (pages 1–84) in which the purposes, chief ideas and the material of the book are made clear, there is a division of the first part under two headings. The business of the pages under the first heading, 'Mathematical Logic', is with the theory of statement connections (pages 90–126), the theory of statements with variables (pages 127–86), and the theory of classes and relations as an algebra (pages 187–326). The business of the pages under the second heading, 'Prolegomena to Cardinal Arithmetic', is with ideas needed for giving a definition of 'cardinal number'

77

and for building an arithmetic of cardinal numbers with the bricks of logic.

The later, second and third parts go into the details of the arithmetics of cardinal and ordinal numbers, these arithmetics being based completely on (what the writers take to be only) logic.

11.2. Relation between Russell's 'Principles' and 'Principia Mathematica'. The writing of *Principia Mathematica* was to take further the teaching of Russell's *Principles of Mathematics*. In the *Principles* Russell had put forward the opinion that all mathematics is—may be got as—the offspring of logic: all the ideas of mathematics may be given definitions using only ideas that are a part of logic, and all the theorems of mathematics may be given demonstrations using only axioms and definitions that are a part of logic. The *Principles* is a long account and discussion of this view in the philosophy of mathematics. But it does not give a detailed development of logic with chains of deductions designed as the support of this view. This detailed development, as far as arithmetic, is offered in *Principia Mathematica* and that was the chief purpose of the writers' ten years' work on the book.

11.3 Russell's way of getting consistency in class theory. In *The Principles of Mathematics* Russell made the attempt to overcome the doubts, whose seed was in the discoveries of Burali-Forti, himself and others, that the algebra of classes and the arithmetic of classes do not have the property of consistency, by ruling that what seem to be well-formed statements, such as '*x* is an element of *x*', but which are the cause of trouble about consistency, are not well-formed; somewhat like 'of is square a 4 root 2', they do not have good sense. A form of this same sort of ruling was taken up again in *Principia Mathematica*, and it has been much talked about as a way to get consistency in the theory of classes. Two sides of the ruling are: firstly, that whatever is about all of a class is not able to be one of the class; and, secondly, that it is not possible for the values of a function to have parts which may be given definitions only in relation to the function itself. An effect of this is that the functions of which a given thing *a* may be an argument (8.5) are unable to be arguments of one another and that they have no things in common with the functions of which they may be arguments. In view of this one is forced to put functions into different levels; starting with *a* and the other things which may be arguments of the same functions as those of which *a* may be an argument, one comes to functions of which *a* is a possible argument, and then to functions of which such functions are possible arguments, and so on.

12

MATHEMATICAL LOGIC AFTER *PRINCIPIA MATHEMATICA*: HILBERT'S METAMATHEMATICS

12.1. The growth of logic after 'Principia Mathematica'. In the years shortly after *Principia Mathematica* came out one was working to make its axiom bases simpler and better. But there has been little agreement with the Frege-Russell-Whitehead opinion

Logic after 'Principia Mathematica':

that mathematics is logic, though some who have been in agreement with it have been high authorities, for example Quine. Outside the algebra and arithmetic of classes as branches of everyday mathematics not formed as axiom systems, the chief business of the logic of mathematics in the last 50 to 60 years has been with the metamathematics of axiom systems of logic and mathematics and with the field of ideas given birth by metamathematics.

12.2. The structure of an axiom system. The metamathematics of a branch of mathematics is generally done in connection with a more or less fully detailed axiom system of that branch. One became conscious of the conditions, and of the need, of a branch of mathematics being put in the form of a more or less fully detailed axiom system only by degrees, the present-day views being mostly the outcome of the writings of Pasch, Peano, Hilbert and others about axiom systems of simple arithmetic or of Euclid's and Lobachevsky's geometries, and of the writings of Frege, Couturat and the 1920's Polish school of logic headed by Lesniewski, Lukasiewicz and Tarski.

In a fully detailed axiom system S, the letters and other signs of the language to be used in S are listed and rules are given saying which complexes of signs are well-formed statements in S; there is a complete list of those well-formed statements which are to be used as axioms; a complete list of used definitions is given; there is an account of what is needed and what is enough for something to be a demonstration, and for something to be a theorem, in S; there is a complete list of the rules of deduction in S, these rules making clear and limiting the moves that may be made in playing with the well-formed statements in S; if the theorems of some other branch of logic or mathematics may be used in S, there is a statement saying which of such branches may be used as helps in S. Roughly, a well-formed statement is a theorem if and only if it is an axiom or a last step in a demonstration, a demonstration being a chain of well-formed statements every one of which is an axiom or rightly got by rules of deduction. If s is any well-formed statement in the language of S, and its opposite, having the sense 's is false', is $-s$, then S is said to have the property of consistency if and only if not the two of s and $-s$ are theorems in S. S is said to be complete if and only if, for any well-formed statement s without free variables, at least one of s and $-s$ is a theorem in S. If S does not have the property of consistency, it will be complete, because every s will become a theorem in S: any well-formed statement may be got as the outcome of two opposite theorems. In all other ways the properties of consistency and being complete are not at all dependent on one another. One's normal purpose in building a fully detailed axiom system S is to get a system that has those two properties.

12.3. Post's, Hilbert and Ackermann's, and Goedel's demonstrations for 'Principia Mathematica' systems.

Emil Post (1897–1954), while still a very young man, gave a demonstration in 1920 that the *Principia Mathematica* axiom system for the logic of statement connections has the property of consistency and that it is complete in the sense that every true law of that logic may be got as a theorem in the system. Being complete in this sense the system is, in addition, complete in the sense given in 12.2. Post's demonstration was based on a decision process that was in all important points the same as Peirce's (7.7). In addition, Post's 1920 paper had an outline account of his invention of an *n*-valued logic of statement connections, that is a logic in which not only the two values *v* and *f* but any number *n* of values may be given to its statements. At the same time the invention of a 3-valued logic was made by Lukasiewicz; for example, if the three values are *v* (true), *f* (false) and *u* (uncertain), then the definitions of the values of N*s* (= 'not-*s*') and of *s* C *t* (= 'if *s*, then *t*') are as in the tables:

s	N*s*
v	*f*
u	*u*
f	*v*

s	*t*	*s* C *t*
v	*v*	*v*
v	*u*	*u*
v	*f*	*f*
u	*v*	*v*
u	*u*	*v*
u	*f*	*u*
f	*v*	*v*
f	*u*	*v*
f	*f*	*v*

The *n*-valued and 3-valued logics of Post and Lukasiewicz are interesting in themselves and are important for certain parts of metamathematics.

Later in the 1920's Hilbert and Ackermann gave a demonstration of the consistency of an axiom system of the logic of statements with variables, a system that was equal to the system in *Principia Mathematica*. And in 1930 Goedel gave a demonstration of the fact that the *Principia Mathematica* system for this logic is complete.

12.4. Brouwer's doubts and Hilbert's metamathematics. The first demonstrations of the consistency of Euclid's geometry, and of Lobachevsky's geo-

Hilbert's Metamathematics

metry, were made in 1899, and in 1903, by the great German authority on mathematics David Hilbert (1862–1943). These demonstrations were made by the help of arithmetic. They have no value if arithmetic is without consistency. This condition of things where the consistency of one system S_1 is dependent on the consistency of another system S_2 is quite general. To give good reasons pointing to the consistency of a system, one has to go outside the system.

Again, no one had been able to get a consistency demonstration for the theory of classes. Because of the ever-present feeling that any branch of mathematics might not have consistency, a feeling made ever-present by the troubles in the theory of classes, the design of getting consistency demonstrations for arithmetic and class theory specially, and for other branches of mathematics, was slowly formed by Hilbert after 1899.

Brouwer (1882–) had, from 1907, been responsible for the suggestion that something is seriously wrong with almost all reasoning in everyday mathematics, in so far as this makes use of classes that are not finite. In relation to such reasoning the 'law' of logic named the *tertium non datur* ('there is no third value (like true, false)') is not true as a general law. The outcome of the belief in this law is that statements about numbers, points, and so on are somehow able to be true without being made, by man's thought, to be true by a building up of examples that make the statements true. But how are numbers, points, and so on to have existence separately from our minds? Does it make sense to have the belief that they have existence till they are given existence by our building operations in mathematics? And if it does not make sense, then it is not possible for statements about the ideas of mathematics, like numbers and points, to be true in themselves, as some strange sorts of facts, and separately from us.

Hilbert made the decision to let only certain sorts of reasonings into his metamathematics, taking note of Brouwer's protests. It was, anyhow, unwise to make use, in a consistency demonstration, of any part of mathematics whose consistency itself was open to some doubt. One effect of this decision was that all Cantor's ordinals that are higher than some quite low level are not to be used in consistency demonstrations.

12.5. Goedel's Theorem in the metamathematics of arithmetic. In the 1920's Hilbert's chief purpose was to get a consistency demonstration of simple arithmetic—of the arithmetic of the natural numbers. However, at the end of the 1920's—in 1931—Goedel (1906–) made a most surprising and shocking discovery of a demonstration of a theorem in the metamathematics of simple arithmetic by which Hilbert's hopes were smashed. Goedel's Theorem itself did not come as a great surprise; there had been earlier suggestions, for example in 1927 by von Neumann in an important paper on metamathematics demonstrations, that something like it was probably true. But experts were surprised that, with the current knowledge, a demonstration of it was able to be made.

Goedel's Theorem is to the effect that no axiom system of simple arithmetic is complete if it has the property of consistency; so whatever axiom system one has of simple arithmetic, with consistency, there is at least one true relation between natural numbers which may not be got as a theorem in it.

12.6. The theory of recursive functions. From the 1930's till now most work in metamathematics has been

Hilbert's Metamathematics

about arithmetic, being guided by the designs of Hilbert and by ideas used by Goedel, and by a number of different forms of these designs and ideas.

Specially important has been the idea of that sort of function which is named *recursive*. Simple functions of this sort have been used for a long time in mathematics; but there was no deep theory of this class of functions till after Goedel's Theorem whose demonstration made great use of recursive functions. A function f is said to be recursive if and only if, for any argument a of the class X of definition of f, X being made from the class N^0 of 0 and the natural numbers, the value b of f for a (8.5) may be fully worked out from a system of equations, the working out being done with the help of other values of f for other arguments and with the help of other functions whose definitions are given by the same system of equations. A very simple example of a recursive function is the power function whose definition is given by the equations (I) $a^0 = 1$, (II) $a^{n+1} = a \times a^n$ where n is an element of N^0. The values of this function for any argument a may be worked out in this way: (1) $a^1 = a \times a^0$ by (II), $= a \times 1$ by (I), $= a$; (2) $a^2 = a \times a^1 = a \times a$, by (1); and so on.

The theory of recursive functions has deep connections with the theory of those functions all of whose values may be worked out by machines. Goedel, Skolem, Church, Kleene and Turing have been chiefly responsible for the development of these new theories. Accounts of them and of the other branches of Mathematical Logic that have been named in our discussion will be given in the other books coming out as *Monographs in Modern Logic*.

FURTHER READING

The purpose of this list is to give the names of some of the chief books on the history of Mathematical Logic. It is not to give the names of the writings of the makers of Mathematical Logic; for these, see Church (3).

1. BOCHENSKI, I. M., *Ancient Formal Logic*, 1951.

2. BOCHENSKI, I. M., *Formal Logic*, translated from the German by Ivo Thomas, 1961.

3. CHURCH, A., 'A Bibliography of Symbolic Logic' *Journal of Symbolic Logic*, 1936, 1938.

4. CHURCH, A. and others, 'Logic, History of', *Encyclopedia Britannica*, Volume 14, 1957.

5. JORGENSEN, J., *A Treatise of Formal Logic, its evolution* . . ., 3 vols., 1931.

6. KNEALE, W. C. and M., *The Development of Logic*, 1962.

7. LEWIS, C. I., *A Survey of Symbolic Logic*, 1918, 1961.

8. LUKASIEWICZ, J., *Aristotle's Syllogistic*, 2nd edition, 1957.

9. MATES, B., *Stoic Logic*, 1953.

10. MOODY, E. A., *Truth and Consequence in Medieval Logic*, 1953.

11. QUINE, W. V. O., 'Whitehead and the rise of modern logic', *Philosophy of Alfred North Whitehead*, ed. by P. A. Schilpp, 1941.

12. SCHOLZ, H., *Abriss der Geschichte der Logik*, 1931.

INDEX

87

Index